TOPICS IN MATHEMATICAL MODELING

TOPICS IN MATHEMATICAL MODELING

K. K. TUNG

PRINCETON UNIVERSITY PRESS
Princeton and Oxford

ISBN-13: 978-0-691-11642-6

This book has been composed in ITC Stone

Printed in the United States of America

To my parents,
David and Lily Tung

Contents

Preface

This book is based on my lecture notes for a college sophomore/junior-level course on mathematical modeling. I designed this course with partial funding from a VIGRE grant by the National Science Foundation to the University of Washington and further developed it into a book when I taught it six times to undergraduate students in our Applied and Computational Mathematical Sciences program, for which this course was a required "overview" of applied mathematics. The style of the book is that of an organized collection of modeling examples with a mathematical theme. The textbook is aimed at students who have had calculus and some ordinary differential equations. Students we surveyed liked seeing how mathematics is applied to real-life situations and were surprised that the simple mathematics they learned could already be useful. However, they overwhelmingly expressed the opinion that they were not interested in the traditional modeling topics such as pendulums and springs, or how long it takes to cook a turkey, a sentiment that I partly share. I also agree that a *deductive* subject such as classical mechanics probably should not belong to a course on *modeling*. To me an important process of modeling is to distill from empirical data a conceptual model and to quantify the conceptual model with a set of governing mathematical equations. To "model" classical mechanics, we probably have to pretend that we are back in the 16th or 17th century. If, however, the topics are broadened from physical sciences to other emerging areas, interesting applied mathematical problems exist that (a) can be captured and solved using simple mathematics and (b) are current research problems whose solutions have important societal impacts. The difficulty is finding such modeling problems and presenting them in a coherent context, while at the same time retaining some historical perspective. In this endeavor I have benefited enormously from my colleagues, who pointed me to interesting research papers in various sources in fields far from my own. I especially would like to thank Professor James D. Murray, FRS, whose encyclopedic knowledge of historical cases and recent developments in mathematical biology is a valuable source of information. So is his book *Mathematical Biology*. Professor Mark Kot taught a similar course, and his lecture notes helped my teaching greatly. Had his spouse agreed to let him write "another book!" he would probably have been my coauthor on this project. His recent monograph, *Elements of Mathematical Ecology*, provides more interesting examples. My former colleague and chairman, Professor

Frederic Wan, started the tradition of teaching a modeling course in our department. I used his textbook, *Mathematical Models and Their Analysis*, for a graduate-level course with the same title. That book has also been used by others at the undergraduate level. At the start of my academic career as an assistant professor, I cotaught a modeling course with University Professor C. C. Lin to Honors sophomores at MIT, using the book by C. C. Lin and L. A. Segal, *Mathematics Applied to Deterministic Problems in the Natural Sciences*, which, in my opinion, was ahead of its time when it was published in 1974. I benefited greatly from all of my colleagues' and teachers' writings and their philosophies. To them I am deeply indebted.

A wonderfully refreshing new book, *Modeling Differential Equations in Biology*, by C. H. Taubes of Harvard, provides a compendium of current research articles in the biology literature and a brief commentary on each. I have found it useful, in teaching our course, to supplement my lecture notes with reading material from Taubes's book when the topics covered are related to biology. On more traditional topics, the 1983 text by Braun, Coleman, and Drew, *Differential Equation Models*, is still a classic. Richard Haberman's *Mathematical Models: Mechanical Vibrations, Population Dynamics and Traffic Flow* became a "Classic" in the SIAM Classics series. Recently I was delighted to come across a new book, *Mathematical Modeling with Case Studies, A Differential Equation Approach Using Maple*, by B. Barnes and G. Fulford. It contains many more nice examples on interacting populations than I have included in the present book.

A comment on pedagogy—although each chapter is independent and can be taught out of order, there is a progression of mathematical themes: from linear to nonlinear problems, from difference to differential equations, from single to coupled equations, and then to partial differential equations. Some chapters are not essential to this progression of mathematical techniques. They can be skipped by an instructor if the focus of the course is more on the mathematical tools. These chapters, however, present interesting phenomena to be modeled and are actually my favorites. Calculus is a prerequisite for this course. If students taking this course are unfamiliar with ordinary differential equations, an instructor may need to spend more time going over "Appendix A: Differential Equations and Their Solutions." Otherwise this appendix (and chapter 3) can be skipped. Students in my class are required to write a term paper to be turned in at the end of the term on a topic of their choice. The paper can be on modeling a phenomenon or a critical review of existing work in the literature. Since this is likely to be the first time a student is writing a scientific report, it is helpful if the instructor gives one lecture on what is expected of them.

This book is based on my lecture notes for a course dealing with *continuous* modeling. Although I have tried to add some *discrete* modeling topics, I probably have not done justice to the diversity of methods and phenomena in that area. As it stands, the book contains more than enough material for a one-semester course on mathematical modeling and gives some introduction to both discrete and continuous topics. The book is written in such a way that there is little cross-referencing of topics across different chapters. Each chapter is independent. An instructor can pick and choose the topics of interest for his/her class to suit the preparation of the students and to fit into a quarter or semester schedule.

I am deeply indebted to Frances Chen, who patiently and expertly typed countless versions of the lecture notes that eventually became this textbook. I am grateful to the students in my modeling class, to the instructors who have taught this class using my lecture notes for correcting my mistakes and typos and giving me feedback, and to William Dickenson for drafting some of the figures. Last but not least are the friendly and professional people I would like to thank at Princeton University Press: senior editor Vickie Kearn, who first contacted me in December 2001 about writing this book and stood by me for the ensuing five years, for her great patience, enthusiasm, and constant encouragement, and her assistant, Sarah Pachner, for her valuable help in obtaining copyright permissions on the figures; and Linny Schenck, my production editor, and Beth Gallagher, my copy editor, for their great work and flexibility. Thank you!

TOPICS IN MATHEMATICAL MODELING

1

Fibonacci Numbers, the Golden Ratio, and Laws of Nature?

Mathematics required:
> high school algebra, geometry, and trigonometry; concept of limits from precalculus

Mathematics introduced:
> difference equations with constant coefficients and their solution; rational approximation to irrational numbers; continued fractions

1.1 Leonardo Fibonacci

Leonardo of Pisa (1175–1250), better known to later Italian mathematicians as Fibonacci (Figure 1.1), was born in Pisa, Italy, and in 1192 went to North Africa (Bugia, Algeria) to live with his father, a customs officer for the Pisan trading colony. His father arranged for the son's instruction in calculational techniques, intending for Leonardo to become a merchant. Leonardo learned the Hindu-Arabic numerals (Figure 1.2) from one of his "excellent" Arab instructors. He further broadened his mathematical horizons on business trips to Egypt, Syria, Greece, Sicily, and Provence. Fibonacci returned to Pisa in 1200 and published a book in 1202 entitled *Liber Abaci* (*Book of the Abacus*), which contains a compilation of mathematics known since the Greeks. The book begins with the first introduction to the Western business world of the decimal number system:

> These are the nine figures of the Indians: 9, 8, 7, 6, 5, 4, 3, 2, 1. With these nine figures, and with the sign 0, which in Arabic is called zephirum, any number can be written, as will be demonstrated.

Since we have ten fingers and ten toes, one may think that there should be nothing more natural than to count in tens, but that was not the case in Europe at the time. Fibonacci himself was doing calculations

Figure 1.1. Statue of Fibonacci in a cemetery in Pisa. (Photograph by Chris Tung.)

Figure 1.2. The Hindu-Arabic numerals.

using the Babylonian system of base 60! (It is not as strange as it seems; the remnant of the sexagesimal system can still be found in our measures of angles and time.)

The third section of *Liber Abaci* contains a puzzle:

> A certain man put a pair of rabbits in a place surrounded on all sides by a wall. How many pairs of rabbits can be produced from that pair in a year if it is supposed that each month each pair begets a new pair which from the second month on becomes productive?

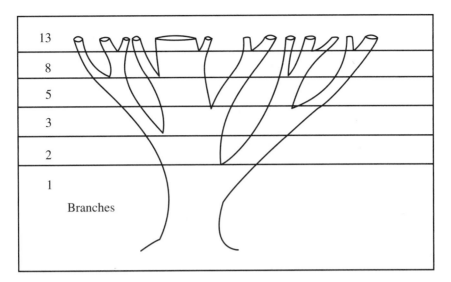

Figure 1.3. Branching of plant every month after a shoot is two months old.

In solving this problem, a sequence of numbers, 1, 1, 2, 3, 5, 8, 13, 21, 34, 55, ..., emerges, as we will show in a moment. This sequence is now known as the Fibonacci sequence.

The above problem involving incestuous rabbits is admittedly unrealistic, but similar problems can be phrased in more plausible contexts: A plant (tree) has to grow two months before it branches, and then it branches every month. The new shoot also has to grow for two months before it branches (see Figure 1.3). The number of branches, including the original trunk, is, if one counts from the bottom in intervals of one month's growth: 1, 1, 2, 3, 5, 8, 13, ... The plant *Achillea ptarmica*, the "sneezewort," is observed to grow in this pattern.

The Fibonacci sequence also appears in the family tree of honey bees. The male bee, called the drone, develops from the unfertilized egg of the queen bee. Other than the queen, female bees do not reproduce. They are the worker bees. Female bees are produced when the queen mates with the drones. The queen bee develops when a female bee is fed the royal jelly, a special form of honey. So a male bee has only one parent, a mother, while a female bee, be it the queen or a worker bee, has both a mother and a father. If we count the number of parents and grandparents and great grandparents, etc., of a male bee, we will get 1, 1, 2, 3, 5, 8, ..., a Fibonacci sequence.

Let's return to the original mathematical problem posed by Fibonacci, which we haven't yet quite solved. We actually want to solve it more generally, to find the number of pairs of rabbits n months after the first pair was introduced. Let this quantity be denoted by F_n.

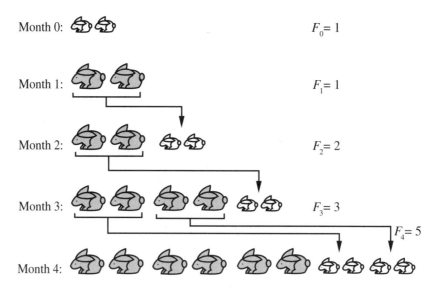

Figure 1.4. Rabbits in the Fibonacci puzzle. The small rabbits are nonproductive; the large rabbits are productive.

We assume that the initial pair of rabbits is one month old and that we count rabbits just before newborns arrive.

One way to proceed is simply to *enumerate*, thus generating a sequence of numbers. Once we have a sufficiently long sequence, we would hopefully be able to see the now famous Fibonacci pattern (Figure 1.4).

After one month, the first pair becomes two months old and is ready to reproduce, but the census is taken before the birth. So $F_1 = 1$, but $F_2 = 2$; by the time they are counted, the newborns are already one month old. The parents are ready to give birth again, but the one-month-old offspring are too young to reproduce. Thus $F_3 = 3$. At the end of three months, both the original pair and its offspring are productive, although the births are counted in the next period. Thus $F_4 = 5$. A month later, an additional pair becomes productive. The three productive pairs add three new pairs of offspring to the population. Thus $F_5 = 8$. At five months, there are five productive pairs: the first-generation parents, four second-generation adults, and one third-generation pair born in the second month. Thus $F_6 = 13$. It now gets more difficult to keep track of all the rabbits, but one can use the aid of a table to keep account of the ages of the offspring. With some difficulty, we obtain the following sequence for the number of rabbit pairs after n months, for $n = 0, 1, 2, 3, 4, 5, 6, 7, 8, 9, 10, 11, 12, \ldots$:
$$1, 1, 2, 3, 5, 8, 13, 21, 34, 55, 89, 144, 233, 377, \ldots.$$

This is the sequence first generated by Fibonacci. The answer to his original question is $F_{12} = 233$.

If we had decided to count rabbits after the newborns arrive instead of before, we would have to deal with three types of rabbits: newborns, one-month-olds, and mature (two-month-old or older) rabbits. In this case, the Fibonacci sequence would have shifted by one, to: 1, 2, 3, 5, 8, 13, 21, 34, 55, 89, 144, 233, ... The initial 1 is missing, which, however, can be added back if we assume that the first pair introduced is newborn. It then takes two months for them to become productive. The discussion below works with either convention.

To find F_n for a general positive integer n, we hope that we can see a pattern in the sequence of numbers already found. A sharp eye can now detect that any number in the sequence is always the sum of the two numbers preceding it. That is,

$$F_{n+2} = F_{n+1} + F_n, \quad \text{for } n = 0, 1, 2, 3, \ldots . \tag{1.1}$$

A second way of arriving at the same recurrence relationship is more preferable, because it does not depend on our ability to detect a pattern from a partial list of answers:

Let $F_n(k)$ be the number of k-month-old rabbit pairs at time n. These will become $(k + 1)$-month-olds at time $n + 1$. So,

$$F_{n+1}(k + 1) = F_n(k).$$

The total number of pairs at time $n + 2$ is equal to the number at $n + 1$ plus the newborn pairs at $n + 2$:

$$F_{n+2} = F_{n+1} + \text{ new births at time } n + 2.$$

The number of new births at $n + 2$ is equal to the number of pairs that are at least one month old at $n + 1$, and so:

$$\begin{aligned} \text{New births at } n + 2 &= F_{n+1}(1) + F_{n+1}(2) + F_{n+1}(3) + F_{n+1}(4) + \cdots \\ &= F_n(0) + F_n(1) + F_n(2) + F_n(3) + \cdots \\ &= F_n. \end{aligned}$$

Therefore,

$$F_{n+2} = F_{n+1} + F_n,$$

which is the same as Eq. (1.1). This recurrence equation is also called the renewal equation. It uses present and past information to predict the future. Mathematically it is a second-order difference equation.

To solve Eq. (1.1), we try, as we generally do for linear difference equations whose coefficients do not depend on n,

$$F_n = \lambda^n,$$

for some as yet undetermined constant λ. When we substitute the trial solution into Eq. (1.1), we get

$$\lambda^{n+2} = \lambda^{n+1} + \lambda^n.$$

Canceling out λ^n, we obtain a quadratic equation,

$$\lambda^2 = \lambda + 1, \tag{1.2}$$

which has two roots (solutions):

$$\lambda_1 = \frac{1}{2}(1 + \sqrt{5}) \text{ and } \lambda_2 = \frac{1}{2}(1 - \sqrt{5}) = -\frac{1}{\lambda_1}.$$

Thus λ_1^n is a solution, and so is λ_2^n. By the principle of linear superposition, the general solution is

$$F_n = a\lambda_1^n + b\lambda_2^n, \tag{1.3}$$

where a and b are arbitrary constants. If you have doubts on the validity of the superposition principle used, I encourage you to plug this general solution back into Eq. (1.1) and see that it satisfies that equation no matter what values of a and b you use. Of course these constants need to be determined by the initial conditions. We need two such auxiliary conditions since we have two unknown constants. They are $F_0 = 1$ and $F_1 = 1$. The first requires that $a + b = 1$, and the second implies that $\lambda_1 a + \lambda_2 b = 1$. Together, they uniquely determine the two constants. Finally, we find:

$$F_n = \frac{1}{\sqrt{5}} \left(\frac{1 + \sqrt{5}}{2} \right)^{n+1} - \frac{1}{\sqrt{5}} \left(\frac{1 - \sqrt{5}}{2} \right)^{n+1}, \quad n = 0, 1, 2, 3, \ldots .$$

$$\tag{1.4}$$

With the irrational number $\sqrt{5}$ in the expression, it is surprising that Eq. (1.4) would always yield whole numbers, $1, 1, 2, 3, 5, 8, 13, \ldots$, when n goes from $0, 1, 2, 3, 4, 5, \ldots$, but you can verify that amazingly it does.

1.2 The Golden Ratio

The number $\lambda_1 = \frac{1}{2}\left(1 + \sqrt{5}\right)$ is known as the *Golden Ratio*. It has also been called the *Golden Section* (in an 1835 book by Martin Ohm) and, since the 16th century, the *Divine Proportion*. It is thought to reflect the ideal proportions of nature and to even possess some mystical powers. It is an irrational number, now denoted by the Greek symbol Φ:

$$\Phi = 1.6180339887\ldots.$$

It does have some very special, though not so mysterious, properties. For example, its square,

$$\Phi^2 = 2.6180339887\ldots,$$

is obtainable by adding 1 to Φ. Its reciprocal,

$$1/\Phi = 0.6180339887\ldots,$$

is the same as subtracting 1 from Φ. These properties are not mysterious at all, if we recall that Φ is a solution of Eq. (1.2).

In terms of Φ, the general solution (1.3) can be written as

$$F_n = a\Phi^n + b\left(-\frac{1}{\Phi}\right)^n.$$

Since $\Phi > 1$, the second term diminishes in importance as n increases, so that for $n \gg 1$,

$$F_n \approx a\Phi^n.$$

Therefore the ratio of successive terms in the Fibonacci sequence approaches the Golden Ratio:

$$\frac{F_{n+1}}{F_n} \to \frac{a\Phi^{n+1}}{a\Phi^n} = \Phi = 1.6180339887\ldots, \text{ as } n \to \infty. \tag{1.5}$$

(In fact, since this property about the ratio converging to the Golden Ratio is independent of a and b, as long as a is not zero, it is satisfied by all solutions to the difference equation (1.1), including the Lucas sequence, which is the sequence of numbers starting with $F_0 = 2$ and $F_1 = 1$: 2, 1, 3, 4, 7, 11, 18, 29, $\ldots$).

For our later use, we also list the result

$$\frac{F_{n+2}}{F_n} \rightarrow \frac{a\Phi^{n+2}}{a\Phi^n} = \Phi^2. \tag{1.6}$$

As you may recall, an irrational number is a number that cannot be expressed as the ratio m/n of two integers, m and n. Mathematicians sometimes are interested in the *rational approximation* of an irrational number; that is, finding two integers, m and n, whose ratio, m/n, gives a good approximation of the irrational number with an error that is as small as possible under some constraints. For example, the irrational number $\pi = 3.14159265\ldots$ can be approximated by the ratio $22/7 = 3.142857\ldots$, with error 0.00126. This is the best rational approximation if n is to be less than 10. When we make m and n larger, the error goes down rapidly. For example, $355/113$ is a rational approximation of π (with n less than 200) with an error of 0.000000266. We measure the degree of irrationality of an irrational number by how slowly the error of its best rational approximation approaches zero when we allow m and n to get bigger and bigger. In this sense π is "not too irrational."

From Eq. (1.5) we see that the value of Φ can thus be approximated by the rational ratio: $8/5 = 1.6$, or $13/8 = 1.625$, or $21/13 = 1.615385\ldots$, or $34/21 = 1.619048\ldots$, or $55/34 = 1.617647\ldots$, or $89/55 = 1.618182\ldots$, or $144/89 = 1.617978\ldots$. The ratios of successive terms in the Fibonacci sequence will eventually converge to the Golden Ratio. One therefore can use the ratio of successive Fibonacci numbers as the rational approximation to the Golden Ratio. Such rational ratios, however, converge to the Golden Ratio extremely slowly. Thus we might say that the Golden Ratio is the *most irrational* of the irrational numbers. (How do we know it is the most irrational of the irrational numbers? A proof requires the use of continued fractions. See exercise 2 for some examples.)

More importantly, the Golden Ratio has its own geometrical significance, first recognized by the Greek mathematicians Pythagoras (560–480 BC), and Euclid (365–325 BC). The Golden Ratio is the only positive number that, when 1 is subtracted from it, equals its reciprocal. Euclid in fact defined it, without using the name Golden Ratio, when he studied the division of a line into what he called the "extreme and mean ratio":

> A straight line is said to have been cut in extreme and mean ratio when, as the whole line is to the greater segment, so is the greater to the lesser.

Figure 1.5. A straight line cut into extreme and mean ratios.

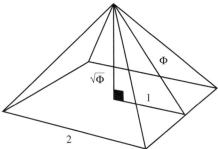

Figure 1.6. The Great Pyramid at Giza and the "Egyptian Triangle."

(Does this sound like Greek to you? If so, you may find Figure 1.5 helpful. Consider the straight line abc, cut into two segments ab and bc, in such a way that the "extreme ratio" $\overline{abc}/\overline{ab}$ is equal to its "mean ratio" $\overline{ab}/\overline{bc}$. Without loss of generality, let the length of small segment $\overline{bc}$ be 1, and $\overline{ab}$ be x, so the whole line $\overline{abc}$ is $1 + x$. The line is said to be cut in extreme and mean ratio when $(1 + x)/x = x/1$; this is the same as $x^2 = x + 1$, which is Eq. (1.2). Φ is the only positive root of that equation.)

Many authors reported that the ancient Egyptians possessed the knowledge of the Golden Ratio even earlier and incorporated it in the geometry of the Great Pyramid of Khufu at Giza, which dates to 2480 BC. Midhat Gazale, who was the president of AT&T-France, wrote in his popular 1999 book, *Gnomon: From Pharaohs to Fractals*:

> It was reported that the Greek historian Herodotus learned from the Egyptian priests that the square of the Great Pyramid's height is equal to the area of its triangular lateral side.

Referring to Figure 1.6, we consider the upright right triangle formed by the height of the pyramid (from its base to its apex), the slanted height of the triangle on its lateral side (the length from the base to the apex of the pyramid along the slanted lateral triangle), and a horizontal line joining these two lines inside the base. We see that if the above statement is true, then the ratio of the hypotenuse to the base of that triangle is equal to the Golden Ratio. (Show this!) However, as pointed out by Mario Livio in his wonderful 2002 book, *The Golden Ratio*, Gazale was repeating an earlier misinterpretation by the English author John Taylor in his 1859 book, *The Great Pyramid: Why Was It Built and Who Built It*, in which Taylor was trying to argue that the construction of the Great Pyramid was through divine intervention. What the Greek historian Herodotus (ca. 485–425 BC) actually said was: "Its base is square, each side is eight plethra long and its height the same." One plethron was 100 Greek feet, approximately 101 English feet (see Fischler, 1979; Markowsky, 1992).

Nevertheless, there is no denying that the physical dimensions of the Great Pyramid as it stands now do give a ratio of hypotenuse to base rather close to the Golden Ratio. The base of the pyramid is approximately a square with sides measuring 756 feet each, and its height is 481 feet. So the base of the upright right triangle is $756/2 = 378$ feet, while the hypotenuse is, by the Pythagorean Theorem, 612 feet. Their ratio is then $612/378 = 1.62$, which is in fact quite close to the Golden Ratio. The debate continues. All we can say is that, casting aside the claims of some religious cults, there is no historical or archeological evidence that the ancient Egyptians knew about the Golden Ratio.

1.3 The Golden Rectangle and Self-Similarity

An application of Euclid's subdivision of a line is to construct a rectangle with the proportion $1 : \Phi$ as the ratio of its short to long side. This rectangle is called the *Golden Rectangle*. Since some of the more familiar proportions in human anatomy, such as the width to the height of an adult face, or the length measured from the top of the head to the navel and from the navel to the bottom of the feet, are *roughly* in the ratio of $1 : \Phi$, speculations abound that artists and sculptors through the ages consciously incorporated the Golden Ratio in their work (see Figure 1.7). (See a critical discussion in Markowsky [1992].)

We shall not be concerned with the subject of the Golden Ratio in art and in defining beauty here. Instead we wish to briefly point out another interesting property of the Golden Rectangle. A Golden Rectangle can be subdivided into a square and a smaller rectangle with the ratio of its short to long side equaling $1/\Phi : 1 = 1 : \Phi$ (since $\Phi - 1 = 1/\Phi$). This is another Golden Rectangle! (See Figure 1.8.) The latter can be subdivided, ad infinitum, into even smaller but similar shapes. The resulting entity is *self-similar*. That is, if you zoom in on a smaller rectangle—with even smaller rectangles and squares embedded in it—and magnify it, it will look the same as the original, bigger rectangle. The property that an object will look the same at all scales is called *self-similarity*. This property is fundamental to the modern concept of *fractals*. (See the exercises for some discussion on fractals.)

Figure 1.7. Leonardo da Vinci's *Mona Lisa*.

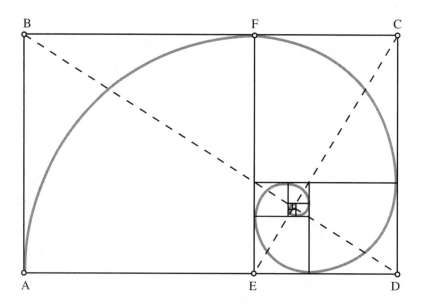

Figure 1.8. Golden Rectangles $AE = AB = 1$, $AD = \Phi$, $FC = \Phi - 1 = 1/\Phi$.

If we join the two opposite corners of each square using a quarter circular arc and connect these arcs together (Figure 1.8), we will obtain a pseudospiral, winding in the limit to a point, which is fancifully called "the eye of God." (The pseudospiral approximates very well a true *logarithmic spiral*, which is also called the *equiangular spiral*. At every point on the logarithmic spiral the angle the tangent makes with the line drawn to the center is always the same.) The logarithmic spiral is self-similar, because it looks the same whatever the magnification. This self-similar property of the spiral may be the reason why some seashells are also in the shape of a logarithmic spiral (Figure 1.9), so as to accommodate the growing body of the mollusk. As it grows, the mollusk constructs larger and larger chambers (and seals off the smaller chambers it no longer uses). Each new chamber has the same familiar shape as the old one the mollusk evacuates.

1.4 Phyllotaxis

Phyllotaxis is the study of leaf arrangements in plants. Fibonacci numbers are found to be "prevalent" in the phyllotaxis of various trees, in seed heads, pinecones, and sunflowers. It is still an ongoing effort by botanists and applied mathematicians to try to understand why this is so from biological and mechanical perspectives.

As the stem of a plant grows upward, leaves sprout to its side, with new leaves above the old ones. How are the new and old leaves

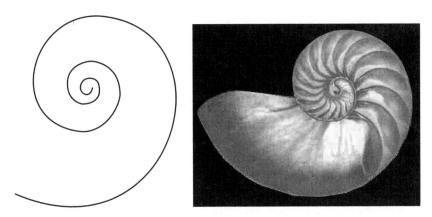

Figure 1.9. The spiral of the Nautilus shell. (Photo courtesy of Bill Strange.)

arranged? Is there a pattern? Theophrastus (372–287 BC) appears to have been the first to notice, in his writing *Enquiries into Plants*, that there is indeed a pattern: "Those that have flat leaves have them in a regular series." The painter Leonardo da Vinci (1452–1519) and the astronomer Johannes Kepler (1571–1630) also wrote on the subject. Kepler in particular noted the connection between the Fibonacci numbers and leaf arrangements. In 1837, the Bravais brothers (Auguste, a crystallographer, and Louis, a botanist) discovered that a new leaf generally advances by the same angle from the previous leaf, and that angle is usually close to 137.5°. That is, if you look down from above on the plant and measure the angle formed between a line drawn from the stem to the leaf and a corresponding line for the next leaf, you will find that there is generally a fixed angle, called the *divergence angle*. You may think that the divergence angle should be something simple, such as 180°, so that the new leaf will be on the opposite side of the stem from the older leaf, "to provide balance" for the plant. This turns out *not* to be advantageous for the plant if it has many leaves, assuming that the sun and rain come from above (vertically). This is because if leaf 0 and leaf 1 are arranged this way, leaf 2 will then be directly above leaf 0, blocking its exposure to sun and moisture.

Generalizing this argument, we note that any divergence angle that is an integer fraction of a circle, i.e., equal to $360°/m$, with m being an integer, is also not optimal for the plant. This is because such an arrangement of leaves is periodic; eventually some new leaves will be directly above some older ones, and the pattern repeats for newer leaves. For example, for $m = 3$, the fourth leaf will be directly above the first. (Similarly, a divergence angle of $360° \, n/m$ will also not be optimal because it is again periodic.) It would appear that the most optimal arrangement would be obtained if we replaced the integer m by an *irrational number*—the more irrational the better. And the best number would seem to be $\Phi = 1.618\ldots$, the Golden Ratio, the most irrational of all irrational numbers. And so, naturally, the *divergence angle* $= 360°/\Phi = 222.5°$, which is the same as $360 - 222.5 = 137.5°$, measuring from the other direction, would appear to be the optimal angle. The divergence angle of $137.5°$ is called the Golden Angle.

Botanists define the *phyllotactic ratio* as the fraction of a circle through which a new leaf turns from the previous (older) leaf. So in this case the phyllotactic ratio is $1/\Phi = 0.618\ldots$. Since this is more than half of a circle, one may want to measure the angle from the other direction (e.g., counterclockwise instead of clockwise). In that case the phyllotactic ratio would be $1 - 1/\Phi = 1/\Phi^2 = 0.382\ldots$. Given the propensity of botanists to list phyllotactic ratios as ratios of integers, it is not surprising that ratios of every other Fibonacci number show up in rational approximation to the phyllotactic ratio of $1/\Phi^2$, in the form of Eq. (1.6). That is,

$$\text{Phyllotactic ratio} \; = \frac{1}{\Phi^2} \approx \frac{F_n}{F_{n+2}}, \tag{1.7}$$

where F_n is one of the Fibonacci numbers. The phyllotactic ratio is the ratio of *every other* Fibonacci number. If one measures the angle in the other direction, e.g., clockwise rather than counterclockwise, one will detect a different set of Fibonacci numbers:

$$\text{Phyllotactic ratio} \; = \frac{1}{\Phi} \approx \frac{F_n}{F_{n+1}}, \tag{1.8}$$

according to Eq. (1.5). This may explain why *three consecutive members* of the Fibonacci sequence are often found in the phyllotactic ratios of a single plant, a situation that may appear mysterious at first sight.

The above argument applies to plants with many, many leaves (in fact, an infinite number of leaves) and under the assumption that the maximum exposure to the sun is the only determining factor for the arrangement of leaves in a plant. Neither of these assumptions is realistic. It remains unexplained why the prevalent tendency is for realistic plants to have a divergence angle close to $137.5°$. Nevertheless, if a plant must choose a *fixed* divergence angle, *why not* choose $137.5°$? There is no *better* angle from which to choose. Now *suppose* a plant of finite height (and with a finite number of leaves) grows leaves at this fixed angle. Then what phyllotactic ratio would be observed? It would, of course, be Eq. (1.7) or (1.8), depending on the direction from which you measure the angle. A plant with a larger number of leaves would generally have a larger value of n, giving a better rational approximation to the Golden Ratio than plants with fewer leaves.

Some examples of phyllotactic ratios for selected plants are given below. A ratio of, e.g., 3/8 means that in three turns of a circle one would find leaf 8 almost directly above leaf 0.

Apple, apricot, cherry, coast live oak, holly, plum	2/5
Pear, weeping willow, poplar	3/8
Pussy willow, almond	5/13

The above explanation of the phyllotactic ratio is somewhat unsatisfactory because we have not explained why the divergence angle is prevalently 137.5°. Surely the preference for maximum exposure to the overhead sun need not be absolute. There must be some other constraints that we have not included in our arguments so far.

A better, though still controversial, argument goes a little deeper in the developmental biology of plants, as we will consider in the next section.

1.5 Pinecones, Sunflowers, and Other Seed Heads

Smith College Botanical Gardens maintains a very informative website on phyllotaxis (http://www.maven.smith.edu/~phyllo/). Recently, two Smith College mathematics professors, Pau Atela and Christophe Golé, along with a colleague, Scott Hotton from Miami University, developed a mathematical model (Atela, Golé, and Hotton, 2002) that can explain the prevalence of a particular divergence angle and the Fibonacci phyllotaxis in seed heads.

A pinecone can be viewed as a "plant" with a very short stem on which many "leaves" (scales) grow, with the newer scales developing near the tip (Figure 1.10). Sunflower heads and other seed heads (Figure 1.11) are extreme versions of such a "plant," where the arrangement becomes two dimensional. The new "leaves" (florets) sprout near the center, and as they grow older and bigger they are displaced radially outward. New florets do not grow on top of the old, because then the old florets would be completely blocked from the sun. Perhaps in these cases, optimizing exposure to the sun may take on more importance than in plants with long stems whose leaves are separated by finite vertical distances. In such tightly packed plants, however, another factor needs to be taken into account, that of the efficiency of packing as the florets or seeds grow.

Let's first state what we would like to explain. First is the divergence angle, of course: How does a plant know to pick 137.5°? Second is

Figure 1.10. A pinecone viewed from the top. In this view, your eyes will pick up 8 counterclockwise spirals and 13 clockwise spirals. (Photo by Rolf Rutishauster, University of Zurich, Switzerland, and used by permission; also courtesy of Atela & Golé, http://math.smith.edu/phyllo.)

a phenomenon that is more apparent in tightly packed seed heads than in leaf arrangements on a branch, and that is the appearance of clockwise and counterclockwise spirals, whose numbers follow the Fibonacci sequence. In Figure 1.10 we show a pinecone. Looking at the pinecone from the top, your eye will pick up spirals. There are actually two sets of spirals in this picture: 8 counterclockwise and 13 clockwise. These spirals are called *parastichies*, and we say that this pinecone has a *parastichy number* of (8, 13). These, mysteriously, are two consecutive numbers in the Fibonacci sequence. Even larger parastichy numbers can be found in sunflower heads. Most common are (34, 55), but larger sunflowers with parastichy numbers of (55, 89), (89, 144), and even (144, 233) have been seen. In a way these are artificial patterns that our eyes pick up; an individual scale (or floret) does not move along such a spiral as it grows away from the center of the stem. It is just that our eyes tend to connect the scales closest to each other

Figure 1.11. The seed head of the coneflower, a member of the daisy family. Note the apparent clockwise and counterclockwise spirals picked up by your eyes. (Photo by Tim Stone and used by permission.)

to form a pattern, and the same scale can be a member of both a counterclockwise spiral and a clockwise spiral. That our eyes will pick up these counter-rotating spirals in consecutive Fibonacci numbers in closely packed points separated by 137.5° was shown in 1907 by the German mathematician G. van Iterson. It is easier for you to show this by *simulation*, i.e., to plot these points on a sheet of paper or generate them on a computer monitor, than it is for us to prove it mathematically. This simulation is done in Figure 1.12, where the florets numbered higher are older.

1.6 The Hofmeister Rule

The patterns we see in large sunflower seed heads are actually already present when the sunflower's blossom is only 2 mm in diameter. In other plants, an electron microscope is needed to see these patterns present in their small shoot tips, called *meristems* (Figure 1.13). A meristem is the growing tip of a plant, which is usually dome shaped. Around the apex of a meristem, cells develop that will later grow to

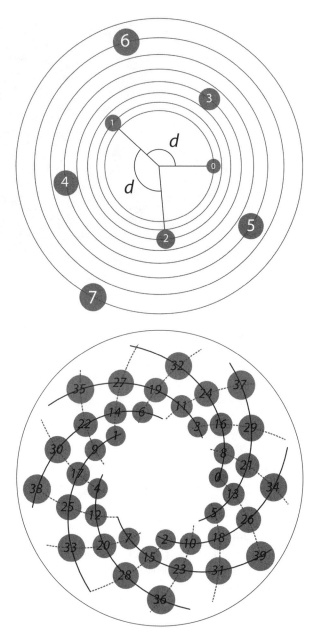

Figure 1.12. In this simulation, points are placed successively at a divergence angle of $d = 137.5°$, then moved radially as they grow larger. When they are packed close together, our eyes pick up counter-rotating spirals: 8 counterclockwise and 13 clockwise in the example shown in the bottom panel. (Courtesy of Atela & Golé, http://math.smith.edu/phyllo.)

Figure 1.13. The meristem of Norway Spruce (top) has spiral parastichy (8, 13). These are primordia of needles. The meristem of artichoke (bottom) has spiral parastichy (34, 55). The primordia are future hairs in the artichoke heart. (Photo of Norway spruce by Rolf Rutishauser, University of Zurich, Switzerland, and used by permission; photo of artichoke courtesy of Jacques Dumais, http://math.smith.edu/phyllo.)

become leaves, petals, or florets. These "embryonic leaves" are called *primordia*. At the stage of their generation, the primordial bulges are crowded together. In 1868, from his microscopic study of plant meristems, the botanist Wilhem Hofmeister proposed that a new primordium always forms in the least crowded spot along the meristem ring. This is now known as the Hofmeister rule. The location of that least crowded spot will depend on how fast the older primordia move away from the apex as they grow. It is not yet agreed how they know to move away "to make room" for new primordia.

1.7 A Dynamical Model

In the mathematical model of Atela, Golé, and Hotton (2002), the essential features of primordial placement and growth are represented by the set of divergence angles $\{d_1, d_2, d_3, \ldots\}$ and the magnification factors on the successive distance of the primordium from the center of the apex, $G > 1$, as it grows away from the apex. (Refer to Figure 1.12, but allow the angles d to be different and arbitrary.)

In a *dynamical* model, a primordium first forms at one point along the edge of the circular apex and then moves radially away from the center as the shoot grows. According to Hofmeister's rule, the next primordium is to be placed farthest away from the first one along the apex ring. It is thus placed 180° away from it, on the other side of the circle. The placement of the third primordium will depend on how fast the first point is moving away from the edge of the apex. In the extreme case, if the first point moves away from the apex very rapidly, the third point should be placed close to where the first point was originally placed, because that is where the third point is farthest away from the second point; the location of the first point is by now too far away to matter. In this case, the divergence angle for the third point is 180°. In this extreme case, the primordia would occupy two radial lines, with a divergence angle of 180°. This yields a parastichy number (1, 1). Something interesting happens when we reduce the growth rate from this extreme value. Then the original position of the first point is no longer necessarily the least crowded spot for the placement of the third point, because now the first point is in the way. The third point should be placed around the apex ring at a location that minimizes its distance from both the first and the second points. This then determines its divergence angle to be somewhere between 180° and 90°. The process is continued for the placement of the fourth point along the apex ring, which has to minimize its distance from all the preceding points. The resulting divergence angle fluctuates a little as more and more points are placed and soon settles down to a *fixed* angle, i.e., $d_1 = d_2 = d_3 = \cdots = d$. As G gets smaller (but still $G > 1$), the divergence angle approaches the Golden Angle: $d = 137.5°$. At the same time as the divergence angle is getting closer to the Golden Angle, the parastichy number first becomes (1, 2) or (2, 1), depending on the direction of the spiral one is following. The branch (1, 2) then becomes (3, 2), then (3, 5), $\ldots$, (F_n, F_{n+1}), as G is slowly reduced, where F_n belongs to the Fibonacci sequence.

The authors point out that when plants make the transition from a vegetative state to a flowering state, the rate at which the primordia are growing apart decreases, and that is when Fibonacci-like parastichy is observed. It therefore appears that the observed appearance of Fibonacci numbers and the Golden Angle may be dictated by the need of the meristem to pack primordia efficiently when they are crowded together.

Recently, a well-known applied mathematician, Professor Alan Newell, and his graduate student Patrick Shipman at the University of Arizona proposed a different model to explain the appearance of Fibonacci numbers in the counter-rotating spirals on plants such as the cactus. Starting with the observation that the spiral patterns are already built into the plant at its earliest developmental stage, and further observing that the tender tip of a growing plant is capped by a thin outer shell, they propose that the spiral pattern is formed as the shell buckles into spiral ridges, so as to relieve mechanical stress. Their mathematical model and analysis appear in the April 23, 2004, issue of *Physical Review Letters*.

1.8 Concluding Remarks

When we observe nature, we often find certain patterns repeating themselves across a wide range of phenomena. Do these patterns reflect the laws of nature? Science would give more credence to those patterns that can be explained by some physically or biologically based mechanisms. A way to test these mechanisms is to incorporate the hypotheses into a mathematical model and then see if the model's predictions agree with observations. It appears that many of the reported sightings of the Golden Ratio in nature may be the result of chance: There are billions and billions of plants and some of them even by random chance would give the appearance of Fibonacci parastichy. However, given a prevalent tendency for plants to follow such a pattern, it is a fruitful area for botanists and mathematicians to build models for the purpose of seeking answers to the question, "Why?"

1.9 Exercises

1. *A puzzle on inheritance*

Fibonacci has this puzzle in *Liber Abaci*: A man whose end was approaching summoned his sons and said, "Divide my money as I shall prescribe." To his eldest son, he said, "You are to have 1 bezant and a seventh of what is left." To his second son, he said, "Take 2 bezants and a seventh of what remains." To the third son, he said, "You are to take 3 bezants and a seventh of what is left." Thus he gave each son 1 bezant more than the previous son and a seventh of what remained, and to the last son all that was left. After following their father's instructions with care, the sons found that they had shared their inheritance equally. How many sons were there, and how large was the estate? Solve this puzzle.

2. *Continued fractions*

a. Show that the Golden Ratio can be expressed in the form of a continued fraction:

$$\Phi = 1 + \cfrac{1}{1 + \cfrac{1}{1 + \cfrac{1}{1 + \cfrac{1}{1 + \cfrac{1}{1 + \ldots \ldots}}}}}$$

Hint: Start with the equation that the Golden Ratio satisfies: $x^2 = x + 1$. Divide by x to get $x = 1 + 1/x$. Substitute $x = 1 + 1/x$ for the x in the denominator, and repeat the process.

b. Show that $\sqrt{2}$ can be written in a continued fraction of the form

$$\sqrt{2} = 1 + \cfrac{1}{2 + \cfrac{1}{2 + \cfrac{1}{2 + \ldots \ldots}}}$$

Hint: Write $\sqrt{2} = 1 + 1/x$ and show that x satisfies $x^2 = 2x + 1$. Divide by x to get $x = 2 + 1/x$. Substitute $x = 2 + 1/x$ for the x in the denominator, and repeat the process.

c. You don't need to do anything for this part. The irrational number $e = 2.71828$ is defined by the limit

$$e = \lim_{n\to\infty} (1 + 1/n)^n.$$

Euler showed that it can be written in the following continued fraction:

$$e = 2 + \cfrac{1}{1 + \cfrac{1}{2 + \cfrac{2}{3 + \cfrac{3}{4 + \cfrac{4}{5 + \cfrac{5}{6 + \ldots\ldots}}}}}}$$

The irrational number π can be written in a continued fraction as (discovered by Brouncker):

$$\pi = 3 + \cfrac{1^2}{6 + \cfrac{3^2}{6 + \cfrac{5^2}{6 + \cfrac{7^2}{6 + \ldots\ldots}}}}$$

d. Truncate each of the continued fractions in (a), (b), and (c) at successive levels to obtain a rational approximation and compare the resulting approximation with the value of the original irrational number. Note and compare the convergence rates of the successive approximations for each irrational number in (a), (b), and (c).

3. **The Hardy-Weinberg law in genetics**

This law concerns the genetic make-up of a population from one generation to the next. It states that in sexually reproducing organisms, in the absence of genetic mutation, factors (called *alleles*) determining inherited traits are passed down unchanged from generation to generation. We want to show that the law is true. Consider the simple case of only two alleles, A and B, in a gene. The probability of occurrence of the A gene in a population in generation n is p_n, and that of the B gene is q_n. $p_n + q_n = 1$. These two alleles combine to form in the next generation AA or BB or AB, with probability p_n^2, q_n^2, and $2p_nq_n$, respectively.

a. The probability of the occurrence of the A alleles in the $n+1$ generation is denoted by p_{n+1} and that of the B alleles by q_{n+1}. We write

$$p_{n+1} = f(p_n, q_n), \quad q_{n+1} = g(p_n, q_n).$$

Find the functions f and g.

Hint: The probability of occurrence of AA in generation $n+1$ from generation n is p_n^2. The probability is 100% that the individual with the AA gene has the A allele. The probability is only 50% that an individual with the AB gene will contribute an A allele to the next generation.

b. Show that $f = p_n$ and $g = q_n$, and therefore

$$p_{n+1} = p_n = p \quad \text{and} \quad q_{n+1} = q_n = q,$$

where p and q are independent of n.

4. *Logarithmic spiral*

Falcons flying to attack their prey on the ground follow a logarithmic spiral instead of a straight line. This is because their eyes are on the two sides of their head, and if they needed to cock their head to keep their prey in their sight it would increase the air resistance. So they fly in a trajectory keeping the same angle, 40°, between its tangent and the direct line to the prey. Show that the requirement of constant angles yields a trajectory that is in the form of a logarithmic spiral, $r = ae^{b\theta}$, where r is the radial distance to the prey and θ is the azimuthal angle.

5. *Fractal dimensions*

A mathematical way to generalize our intuitive way of defining the number of dimensions is to define it as the scaling exponent d. If you have a square, a two-dimensional object, and you divide its length and width by a scaling factor—say 2—what you obtain is four smaller squares. So, $4 = 2^d$. Solving, we get $d = 2$, and we conclude that the dimension of the square is 2. If you divide a line into 2, you will get two shorter lines: $2 = 2^d$. So the dimension of the line is 1. Similarly $d = 3$ for a cube. Now consider the fractal shape called the Sierpinski triangle (Figure 1.14). It is constructed in the following way: Start with an equilateral triangle of solid color. Connect the midpoints of each of its sides by a straight line. Take out the inverted triangle in the middle

Figure 1.14. The Sierpinski triangle. (Bottom image courtesy of Anthony W. Knapp.)

thus formed. We continue this process with the solid small triangles, until we get the bottom figure in Figure 1.14. What is its dimension? Determine it in the following way:

Divide each of its lengths by 2. How many smaller Sierpinski triangles do you now have contained in the original larger triangle? Set that number to 2^d. Determine its dimension d.

6. *Two-dimensional fractal lung and the Golden Tree*

The blood vessels in a lung branch into ever smaller capillary vessels to facilitate the exchange of oxygen with the blood carried by the capillary vessels. One hypothesis is that the resulting form of the branching maximizes the surface area covered by the vessels. We can test this hypothesis with a very simple (and admittedly not too realistic) model of a "lung" in two dimensions. Consider a blood vessel of unit length, which branches into two vessels each of length f, with $f < 1$. The two smaller vessels are 120° apart. Each of them then branches in the same way, with the new branches reduced in length by the same factor f. Repeat this process ad infinitum. We obtain then a fractal tree, which is called the *Golden Tree*. For too small a reduction factor, say $f = 0.5$, for example, you will find that large gaps of space remain that are not covered by the branches of the tree. With too large a reduction factor, say $f = 0.7$, you will find that the branches overlap. Find the optimal f that yields an arrangement with the branches just about to touch. This optimal f turns out to be $1/\Phi$.

Hint: Assume without loss of generality that the original stem of length unity is oriented vertically, and branching occurs at the top point of the stem. By graphing the tree, you will notice that because of symmetry it suffices to find the condition for the main branches to touch horizontally. They in fact touch at a point directly above the first stem. The condition for touching is that the sum of the horizontal projections of all branches with decreasing lengths starting with the branch of length f^3 be equal to the horizontal projection of the large branch of length f. That is,

$$f \cos 30° = f^3 \cos 30° + f^4 \cos 30° + f^5 \cos 30° + \cdots .$$

Canceling the cosine factor and noting that the right-hand side is a geometric series that you can sum up exactly, you will find an algebraic equation for f.

2

Scaling Laws of Life, the Internet, and Social Networks

Mathematics required:
> logarithms, log-log plots, geometric series, high school probability, simple binomial expansion

Mathematics introduced:
> self-similarity; first-order difference equation with variable coefficient and its solution by iteration

2.1 Introduction

Biology, the study of life forms, has historically been an empirical field, relying mainly on observations and classifications, but progress in the last few decades is beginning to transform it rapidly from a qualitative to a quantitative science. Mathematical modeling is playing an important role in this process. In this chapter we discuss the mathematics of networks, examples of which are the vascular blood network, which brings oxygen to the tissues served by the capillaries, and the plant vascular network, which transports nutrients from roots to leaves.

The World Wide Web has in recent decades been developed exponentially in size and complexity by random individuals attaching their web pages to sites of their choice, apparently without an overall design. Yet, when examined, the huge creature created in this way appears to contain structures in common with biological organisms. Advances in network theory have now allowed a study of these and other very complex networks, including social networks of people (such as actors) and the network formed by author citations. Hopefully these results will one day help us understand the even more complex neural networks in our brain.

2.2 Law of Quarter Powers

An important empirical law in biology is the allometric scaling law, which tells how an animal's property scales with its size or mass. (*Allo* comes from the Greek word *allos*, meaning "other." So allometric

Figure 2.1. Professor Max Kleiber (1893–1976)
of the University of California at Davis, Department of Animal Husbandry.

means "other than metric," or other than linear.) Although size does
matter, we don't live in a world where, e.g., our strength, as measured
by the weight we can lift, varies ("scales") linearly with our own weight.
If this were the case, we would be able to lift 50 times our body weight,
as some ants are able to do. (Presumably, the *isometric* scaling on Planet
Krypton is what endows Superman with his superhuman strength, or
so reasoned the creator of this character.) Or, the fact that a cat is 100
times more massive than a mouse would mean that the cat would live
100 times as long. These we know not to be true. If biology does not scale
linearly with size, how does it scale?

In 1932, Max Kleiber (Figure 2.1) plotted the logarithm of mass (M) in
kilograms of various animals and birds on one axis and the logarithm
of their basal (resting) metabolic rate (Y) (the amount of calories they
consume each day) in kcal/day on another axis, and beheld an amazing
sight. Figure 2.2 is a recent reproduction of Kleiber's plot by West and
Brown (2004) in *Physics Today*. From dove to hen, from rat to man, from
cow to steer, across four orders of magnitude (a factor of 10^4) difference

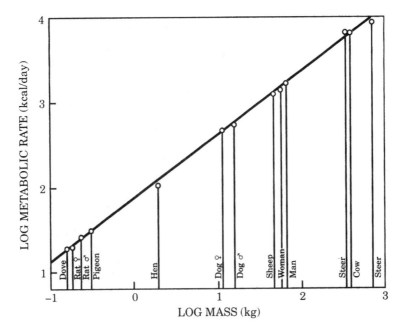

Figure 2.2. The basal metabolic rate of mammals and birds was originally plotted by Max Kleiber in 1932. In this reconstruction, the slope of the best straight-line fit is 0.74, illustrating the scaling of metabolic rate with the $\frac{3}{4}$ power of mass. The diameters of the circles represent estimated data errors of $\pm 10\%$. We use the notation of log to denote logarithm of base 10, so that $\log(10^a) = a$. Taken from West and Brown, *Physics Today*, September 2004.

in mass, the data lie on a single curve—a straight line in fact! The slope of this straight line is $b \sim 0.74$, or about $\frac{3}{4}$.

Mathematically, Kleiber showed that

$$\log Y = b \log M + C.$$

This implies that

$$Y = Y_0 M^b \qquad (2.1)$$

for some constant $Y_0 = 10^C$ (which may be different for different taxonomic groups of organisms), implying that there is a power-law dependence on mass M. Because the exponent b is less than 1, larger animals tend to have slower metabolic rates. A mouse must eat half its body weight each day so as not to starve to death, while the human's daily consumption is only 2% of his or her body weight. The difficult part is to come up with a theory for why the power (exponent) is a simple multiple of a quarter, i.e., $b \sim \frac{3}{4}$. This "law of quarters" went unexplained for 50 years.

In 1883, a German physiologist, Max Rubner, actually proposed a scaling law and explained it thus: if an animal is L times taller than another, then its surface area should be L^2 greater and its mass L^3 greater, since mass is proportional to volume. Its metabolic rate, then, which depends on the amount of heat it sheds, should vary according to its surface area, L^2, which is proportional to $M^{2/3}$. It at least explains why the power b should be less than 1; otherwise a mouse scaled to the size of a cow linearly would burn itself to death by the amount of heat its large body would now generate but that its surface would be unable to dissipate. Unfortunately this law of thirds did not hold up, and Kleiber appeared to have proved it wrong, although this matter is still being debated. In his book on biological scaling, Schmidt-Nielsen (1984) concluded that "the slope of the metabolic regression line for mammals is 0.75 or very close to it, and definitely not 0.67."

There are many such scaling laws with the mysterious quarter powers. The lifespan of mammals was found to scale with their mass as $M^{1/4}$. Larger animals tend to live longer, but not as long as they would if the lifespan scaled linearly with their size. Larger mammals also have slower heart rates. Their heart rates scale with their mass as $M^{-1/4}$. So, the product of lifespan and heart rate, which is the total number of heartbeats in an animal's lifetime, is independent of the size of the animal. It is about 1.5 billion heartbeats. That is how many heartbeats each of us has in total. It seems that when we use it up, we die. A mouse just uses it up faster than a cat. A cat, which is 100 times more massive than a mouse, then lives $(100)^{1/4} \sim 3$ times as long as a mouse.

2.3 A Model of Branching Vascular Networks

Professor James H. Brown is a field animal biologist at the University of New Mexico. He wrote on this subject and lectured in his class about the quarter-power scaling law for animals of various sizes. His graduate student Brian Enquist wondered if the same law applied to plants. For his PhD dissertation, Enquist discovered the equivalent of Kleiber's law for plants. So, there seems to be a universal scaling law, whether or not the organism has a heartbeat.

Brown and Enquist suspected that the vascular networks that transport nutrients in plants and oxygen in animals share some commonality in the way they scale with size. They realized, however, that they needed more mathematical power in order to solve this problem. In 1995, Brown and Enquist got together with Geoffrey West, a physicist

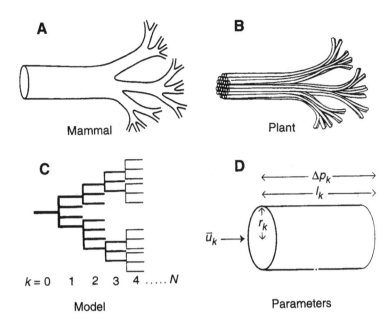

Figure 2.3. (A) Mammalian circulatory and respiratory systems composed of branching tubes; (B) plant vessel-bundle vascular system composed of diverging vessel elements; (C) a schematic representation of such networks, where k specifies the order of the level, beginning with the aorta ($k = 0$) and ending with the capillary ($k = N$); and (D) parameters of a typical tube at the kth level. Taken from West, Brown, and Enquist (1997).

and mathematician at the Santa Fe Institute in New Mexico. Their groundbreaking paper on "A General Model for the Origin of Allometric Scaling Laws in Biology" was published in the leading journal *Science* in 1997 (West, Brown, and Enquist, 1997). The mathematics used in that paper is more advanced than the prerequisites for this book. Here we present a simplified derivation. It should be kept in mind that this model still has some loose ends and is still controversial.

The trio focused first on what all animals and plants have in common: a branching vascular network for distributing nutrients or oxygen. In Figure 2.3 we show one such network for a mammal in (A) and for a plant in (B). In humans, our vascular network transports blood with oxygen in a branching network starting with the aorta ($k = 0$) and ending in the capillary ($k = N$) in $N \sim 22$ levels, with a branching ratio of about $n \sim 3$ (the number of daughter branches arising from one parent branch). At the capillary level, the nutrients are delivered to the cell tissues serviced by each capillary and wastes are collected.

The authors proposed that the properties of the smallest units in such a network, the capillaries, are the same for animals of different sizes. Thus the capillaries of an elephant have the same width as those of a mouse, under this assumption. Animals of smaller sizes than humans have fewer branching levels (N smaller than 22), while larger-sized animals have even more branching levels. As a result of natural selection, a similar design is used at different scales. This is called *self-similarity*. A self-similar picture is one in which, if you zoom in on a smaller portion of the picture and magnify it, it will look just like the bigger picture. West, Brown, and Enquist called this property "self-similarity fractal" and they used the phrases "self-similar" and "fractal" interchangeably, although strictly speaking what they had in mind was only "fractal-like." To be truly "fractal" the self-similarity should be extendable to smaller and smaller scales, and not terminate at the capillary level.

Specifically, the self-similarity assumption says that the ratio of the widths of the vessel from one level to the other:

$$\beta_k \equiv r_{k+1}/r_k = \beta,$$

the ratio of their lengths:

$$\gamma_k \equiv l_{k+1}/l_k = \gamma,$$

and the ratio of their number of branches from one level to the other:

$$n_k = n,$$

are all independent of k. These "fractal properties," as the authors called them, are in fact related. As one vessel branches into n daughter vessels, the cross-sectional area should be preserved to allow a smooth fluid flow. This is called the *area-preserving principle*. Therefore:

$$\pi r_k^2 = n\pi r_{k+1}^2. \tag{2.2}$$

This then implies that

$$r_{k+1}^2/r_k^2 = n^{-1},$$

or

$$\boxed{\beta = n^{-1/2}.} \tag{2.3}$$

An additional assumption is called the *volume-filling principle* and is somewhat more difficult to understand because of the ambiguous way it was stated by West et al. This led to some controversy in the scientific literature later (see exercise 2). Essentially, this principle ensures that all cells in a body are serviced by the capillaries. A single capillary of length l_N and radius r_N services the surrounding cells of volume $\pi \rho^2 l_N$, where $\rho \gg r_N$ is the radius of this cell-cylindrical volume surrounding a single capillary vessel and receiving nutrients from it. It is assumed that

$$\rho \propto l_N.$$

Thus the total volume V serviced by all the capillaries (whose number is N_N in their clumsy notation) is

$$V = \pi \rho^2 l_N N_N \approx C l_N^3 N_N, \tag{2.4}$$

where C is a constant of proportionality. This total volume is the volume of the body of the animal. This same body is also serviced by the vessels in the next level up from the capillary. So it then follows that

$$C l_k^3 N_k \approx C l_{k+1}^3 N_{k+1}$$

(as long as k is large). From it one obtains

$$l_{k+1}^3 / l_k^3 \approx N_k / N_{k+1} = n^{-1}.$$

It then follows that

$$\boxed{\gamma \approx n^{-1/3}.} \tag{2.5}$$

Now consider the total volume of blood contained in all the vessels:

$$V_{\text{blood}} = N_0 V_0 + N_1 V_1 + \cdots + N_N V_N,$$

where $V_k = \pi r_k^2 l_k$ is the volume contained in each blood vessel at level k. The number of blood vessels at level k is $N_k = n^k$. Using self-similarity, this sum can be written in the form

$$V_{\text{blood}} = \pi [r_0^2 l_0 + n r_1^2 l_1 + \cdots + n^N r_N^2 l_N]$$
$$= V_0 [1 + (n\beta^2 \gamma) + \cdots + (n\beta^2 \gamma)^N].$$

This is a *geometric series* in the form of

$$S = 1 + r + r^2 + r^3 + \cdots + r^N,$$

where each succeeding term in the sum is a constant factor r of the term preceding it. If you have forgotten how to sum a geometric series, here is the trick. Multiply S by r:

$$rS = r + r^2 + \cdots + r^N + r^{N+1}.$$

Subtract it from S and observe that all the middle terms cancel:

$$S - rS = 1 - r^{N+1}.$$

Thus the sum of the geometric series is obtained as

$$S = \frac{1 - r^{N+1}}{1 - r}.$$

Using this formula, we can now sum up the blood volumes in the vascular network:

$$V_{\text{blood}} = V_0 \frac{1 - (n\beta^2\gamma)^{N+1}}{1 - (n\beta^2\gamma)}.$$

Because $(n\beta^2\gamma) = n^{-1/3}$ is less than 1, raised to a large power $(N+1)$ it will be reduced to a very small number. Therefore, approximately:

$$V_{\text{blood}} \cong V_0/[1 - (n\beta^2\gamma)]. \tag{2.6}$$

Since the volume of one capillary vessel is

$$V_N = \pi r_N^2 l_N = V_0(\beta^2\gamma)^N,$$

we can rewrite (2.6) as

$$V_{\text{blood}} \cong V_N \frac{(\beta^2\gamma)^{-N}}{1 - (n\beta^2\gamma)}. \tag{2.7}$$

2.4 Predictions of the Model

The total volume of blood in an animal scales with its size and hence mass M, while the volume of one capillary is independent of size by our prior assumption that the smallest units in our network, the capillaries, are invariant. Equation (2.7) implies that

$$(\beta^2 \gamma)^{-N} \propto M$$

or

$$(\beta^2 \gamma)^{-N} = M/M_0$$

for some constant of proportionality $1/M_0$. Taking the log on both sides, we find that, using Eqs. (2.3) and (2.5):

$$N = -\frac{\log(M/M_0)}{\log(\beta^2 \gamma)} = (3/4)\frac{\log(M/M_0)}{\log(n)}. \tag{2.8}$$

The fact that the total number of branches in an animal is proportional only to the logarithm of its mass implies that this fractal-like design is quite efficient for the larger animals. A whale is 10^7 times more massive than a mouse but has only 7 times more branchings from aorta to capillary.

The total number of capillaries is given by

$$N_N = n^N = (M/M_0)^{3/4}. \tag{2.9}$$

The metabolic rate Y is proportional to the rate of flow of nutrients through the vascular network, Q_0, where if we let Q_k be the fluid flow rate through each kth-level vessel, then Q_0, being that through the aorta, would be the total fluid flow rate. The total flow rate can also be calculated at each level, and it should be Q_k times the total number of such vessels, N_k. Thus conservation of fluid flow through each level implies

$$Q_0 = N_k Q_k = N_k \pi r_k^2 u_k = N_N \pi r_N^2 u_N, \tag{2.10}$$

where u_k is the mean velocity of the fluid through the kth-level vascular vessel. Given our scaling for N_k and r_k, (2.10) implies that the fluid velocity is constant through the vascular network all the way to the capillary, and since the latter is invariant with respect to size, the fluid velocity with which the nutrient is delivered is also invariant. Equations (2.10) and (2.9) imply

$$Y \propto Q_0 \propto N_N = (M/M_0)^{3/4}. \qquad (2.11)$$

We have thus derived Kleiber's $\frac{3}{4}$ power law (2.1) for the metabolic rate.

Equation (2.10) also implies that the radius of the aorta vessel, r_0, should scale with mass as $M^{3/8}$ (exercise 1), very close to the observed scaling of 0.36. The length of the aorta, l_0, should scale as $M^{1/4}$.

Another consequence of this scaling with size is explored in West, Brown, and Enquist (2001). They found that if time is normalized by the quarter power of the mass of the mature animal, its growth in time for different animals, be they mammals, birds, fish, or crustaceans, is described by the same universal curve, as seen in Figure 2.4. A derivation of this "universal law of growth" is left to exercise 4. In particular, it is shown that the interval between heartbeats should scale as $M^{1/4}$.

For plants, the rate of resource used by each plant (its metabolic rate Y) is dependent on its size as $M^{3/4}$. Therefore, in an environment with a fixed supply of resources, the maximum number N_{max} of plants of average size M must scale as $M^{-3/4}$. Plant ecologists commonly use M as the dependent variable, and thus

$$M \propto N_{max}^{-4/3}.$$

This is the so-called $-\frac{4}{3}$ law in plant ecology.

2.5 Complications and Modifications

The *area-preserving principle* of (2.2) is more valid for plant vascular systems than it is for animals with a beating heart. The terminal units of a plant's vascular network are the leaves. It is the transpiration at the leaves that creates the vapor pressure that sucks nutrients through its fiber vessel from the roots, up the trunk and branches, and eventually to the leaves. This "pumping" process is helped along by the osmotic pressure. The vessel tubes are of approximately the same diameter

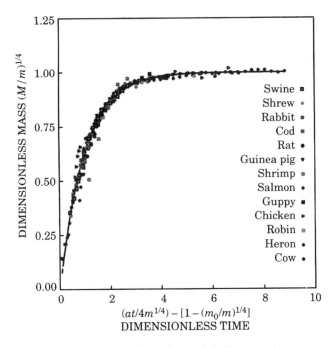

Figure 2.4. The universality of growth is illustrated by plotting a dimensionless mass variable against a dimensionless time variable. Data for mammals, birds, fish, and crustacea all lie on a single universal curve. The quantity M is the mass of the organism at age t, m_0 its birth mass, m its mature mass, and a a parameter determined by theory in terms of basic cellular properties that can be measured independently of growth data. Taken from West and Brown, *Physics Today*, September 2004.

throughout this vascular network, but they are bundled together to form the vessel bundles, which then split up into smaller and smaller bundles from the trunk (the "aorta") to the branches (the "arteries"). Cross-sectional area is preserved in this arrangement (see Figure 2.3B). A consequence of the area-preserving principle is the constancy of the fluid velocity within the vessels (from Eq. (2.10)). For the mammalian vascular network, two complications arise. First, the flow, pumped by a beating heart, is pulsatile. Second, as the branching proceeds from large to small tubes, from aorta to large arteries to smaller arteries, viscosity becomes important, and the flow slows down, which is inconsistent with the area-preserving principle. West, Brown, and Enquist (1997) incorporated these complications and showed that the scaling exponent is dominated by the larger arteries, which are less affected by viscosity

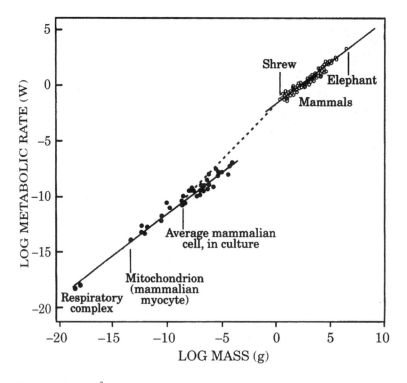

Figure 2.5. The $\frac{3}{4}$ power law for the metabolic rate as a function of mass is observed over 27 orders of magnitude. The masses covered in this plot range from those of individual mammals, to unicellular organisms, to uncoupled mammalian cells, mitochondria, and terminal oxidase molecules of the respiratory complex. The solid lines indicate $\frac{3}{4}$ power scaling. The dashed line is a linear extrapolation that extends to masses below that of the shrew, the lightest mammal. Taken from West and Brown, *Physics Today*, September 2004.

and therefore still satisfy the area-preserving principle. Therefore, the previously obtained $\frac{3}{4}$ power law still holds approximately. For small mammals, such as a 3-gram shrew, the viscous flow starts to dominate just beyond the aorta. There is some indication that smaller mammals deviate from the $\frac{3}{4}$ scaling, possibly for this reason.

2.6 The Fourth Fractal Dimension of Life

Amazingly, the $\frac{3}{4}$ power law holds not only for plants and animals that have a branching vascular network to carry nutrients but also for single-cell (called unicellular) organisms as well (see Figure 2.5). Thus the $\frac{3}{4}$ power scaling law appears to span 27 orders of magnitude in mass.

What is the basis for this "universal scaling law of life"? Certainly the model of branching networks needs to be generalized. The *Science* paper by West, Brown, and Enquist (1999) attempts such a generalization by using concepts from fractal geometry and general but less mechanism-specific statements such as, "Natural selection has tended to maximize both metabolic capacity, by maximizing the scaling of exchange surface areas, and internal efficiency, by minimizing the scaling of transport distances and times. These design principles are independent of detailed dynamics and explicit models and should apply to virtually all organisms." And, "Fractal-like networks effectively endow life with an additional fourth spatial dimension." The basic idea behind this proposal is that the surface that matters is not the smooth skin enclosing the exterior, as in Max Rubner's 1883 argument, but the internal surface across which exchange of nutrients takes place. Evolution has selected organisms that maximize the internal surface inside a compact external volume. It leads to the internal surfaces filling up the volume, endowing it with apparently more of the dimensions of a volume (3) than the dimensions of the exterior surface (2). This idea has not yet been commonly accepted, and we await further development in the field. Not wanting to draw too much attention to it, I have left the derivation of this result to exercise 5, where the idea of a fractal dimension is also discussed.

2.7 Zipf's Law of Human Language, of the Size of Cities, and Email

Zipf's law is named after the Harvard linguistics professor George Kingley Zipf (1902–1950). Zipf was interested in uncovering the fundamental law of human language. He was independently wealthy and used his own money to hire a roomful of human "computers," who would count the number of times an English word, such as "and" or "the," occurs in a given text. He found a power law relating the frequency of occurrence of a word and the rank of that word. The power law is of the form of Eq. (2.1), with the exponent equal to approximately -1. That is, let k be the rank of a word. For example, in most English texts, the word "the" is the most often used and therefore is given the rank $k = 1$. The second most used word is probably "of," and it is given the rank $k = 2$, etc. Let f_k be the frequency of occurrence of each word of rank k, i.e., how many times this word appears in a given text. Zipf's law then says that

$$f_k \propto k^{-b} \tag{2.12}$$

with $b \sim 1$.

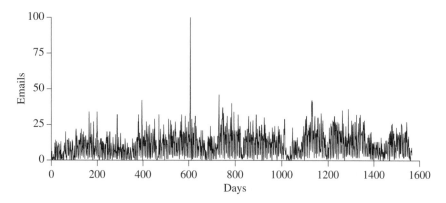

Figure 2.6. The number of emails Professor Mark Kot received each day.

If Zipf's law holds for these k's then "of" will appear half as often as "the." This is approximately true in the *Brown Corpus of Standard American English*, a compilation of American English from various sources of one million words, where "the" occurs 7% of the time and "of" 3.5% of the time. For different texts the rank of a particular word may be different. Analyzing the different Zipf curves for texts written by different people may help identify authorship and uncover evidence of plagiarism.

Zipf also analyzed the sizes of cities, as measured by their populations, plotted them against the ranks of the cities, and showed that they also satisfy a power law with $b \sim 1$. In the United States, the largest city is New York City, with rank $k = 1$, followed by Los Angeles with $k = 2$ and Chicago with $k = 3$, and so on. Plotting the size against rank will also yield a Zipf curve of the form of (2.12). This law appears to apply to most developed countries, but not so well to countries with unique political, economic, or cultural constraints on the movement of its citizens.

My colleague Professor Mark Kot noticed that the number of emails he has received over the past few years appears to follow Zipf's law as well. Figure 2.6 shows the number of emails he received each day. As is typical of most academics, Kot's email number seems to follow an academic-year cycle, with a pause every seven days (on Sunday). When plotted in a log-log plot against the rank of the sender, it approximates a straight line with slope -1 (Figure 2.7). At the low end of the curve, there was a large number of senders who sent only one email, and a smaller number who sent two pieces. The sender ranked number 1, represented by a dot on the high end of the curve, was me. Professor Kot remarked that although I sent the most email to him, I did not send as much as I should have to satisfy Zipf's law. Overall, however, Zipf's law appears to be approximately true here, and we would expect that with

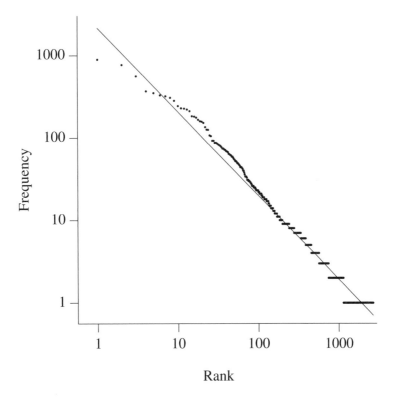

Figure 2.7. Log-log plot of the number of emails against the rank of the sender.

more data, the resulting curve should look better because it may better average out a few individuals' particular behaviors.

Kot, together with his biologist wife, Celeste Berg, and his former PhD student Emily Silverman, looked at the postings in several open biology newsgroups. Submissions (postings) to two such newsgroups are shown in Figure 2.8 as a function of rank of the sender. An almost straight line with a negative slope b is found, implying a power law in the form of (2.12), though b can be different from 1. Their finding, together with an explanation of the observed results, is published in Kot, Berg, and Silverman (2003).

There are various proposed models of varying degrees of complexity that try to explain these scaling laws. For example, Marsili and Zhang (1998) showed that Zipf's law for city distribution can be explained by assuming that citizens interact with each other in a pairwise fashion in choosing a city in which to reside. Kot, Berg, and Silverman (2003) proposed a stochastic (probabilistic) model for the scaling law of newsgroups and explained it by computer simulation. Underlying both models is the phenomenon that "the rich get richer."

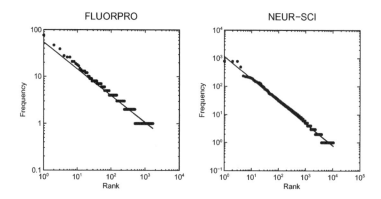

Figure 2.8. Log-log plot of the frequency of submissions as a function of the rank of the sender for two newsgroups, FLUORPRO and NEUR-SCI. Taken from Kot, Berg, and Silverman (2003).

On the other hand, Zipf's original law on word frequency, which he thought revealed the deep law of human language, turns out to be rather "shallow." Wentian Li of the Santa Fe Institute showed in 1992 that a monkey randomly hitting keys on a typewriter with M letters and 1 blank space generates a random text whose word frequency vs. rank obeys Eq. (2.12), with the exponent b given by $\ln(M+1)/\ln(M) \sim 1$, with weak dependence on M as long as $M > 1$ ($b = 1.01158$ for $M = 26$ in English). The derivation is left to exercise 6.

2.8 The World Wide Web and the Actor's Network

In 1998, physicist Albert-László Barabási and his colleagues Hawoong Jeong and Reka Albert at the University of Notre Dame embarked on a project to map the World Wide Web, using a virtual robot to hop from one web page to another and collect the links to and from each web page. Counting how many web pages have k links, they were surprised to find that while more than 80% of them have fewer than four links, 0.01% of them have more than a thousand. Some even have millions. The probability (or the fraction) that any node is connected to k other nodes was found to be proportional to $k^{-\gamma}$, a power law. When plotted on a log-log scale, it follows a straight line with a negative slope of about $\gamma = 2.1$ (see Figure 2.9B). Similar power laws were found in large social networks, such as the network of actors, where each actor represents a node and two actors are linked if they were cast in the same movie together. The negative slope in this case is $\gamma = 2.3$ (see Figure 2.9A). For some unexplained reason, the slope seems to lie between 2 and 3 for large networks. The smaller network of power grids in the United States

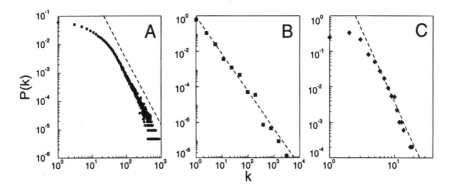

Figure 2.9. The distribution function of connectivities for various large networks. (A) Actor collaboration graph with $N = 212{,}250$ vertices, and average connectivity $\langle k \rangle = 28.78$. (B) World Wide Web, $N = 325{,}729$, $\langle k \rangle = 5.46$. (C) Power-grid data, $N = 4{,}941$, $\langle k \rangle = 2.67$. The dashed lines have slopes (A) $\gamma_{\text{actor}} = 2.3$, (B) $\gamma_{\text{www}} = 2.1$, and (C) $\gamma_{\text{power}} = 4$. From Barabási and Albert (1999).

(see Figure 2.9C) follows a power law with negative slope of $\gamma = 4$. These results were published in *Science* in 1999. That paper, by Barabási and Albert (1999), became highly influential.

The authors were surprised by their finding because they thought people would have different reasons for deciding which sites to link their own web pages to, depending perhaps on the subject matter of their page. Given the diversity of subjects and interests and the vast number of sites one could link to, the situation should appear fairly random. Yet the authors' finding contrasts with the prediction of the standard theory of random networks, which proved that given a number of nodes and random connections between them, the resulting system will have a deeply democratic feature: most nodes will have approximately the same number of links. In the case of power laws, there is no typical number of links, and any number of links is possible. The authors called this type of network "scale free" since it does not have a typical number of links.

There are two important differences between a real network such as the World Wide Web and the textbook case of random networks. The real network is growing, by the continual addition of nodes, and the new nodes preferentially attach to sites that are already well connected. So again the rich get richer, resulting in a few sites (such as Google and Yahoo) with a huge number of links. Early versions of the model concentrated on the development of "scale free" networks, i.e., on explaining why the dependence on the number of links is a power law. Getting the exact value of the exponent was not a high priority, and the value of the exponent predicted, 3, was deemed good enough. This is in contrast to the case of West, Brown, and Enquist's model of allometric scaling in biological organisms, where the central problem is predicting the exact exponent. Later versions of the network model included more complicated processes of attachment to nodes to get a more specific exponent. We will give here a simplified derivation of the power law, following the earlier model of Price (1976), which preceded that of Barabási and Albert by two decades, although the Price model was about paper citations in journals. A good review can be found in Newman (2003).

2.9 Mathematical Modeling of Citation Network and the Web

Derek de Solla Price (1922–1983) was a physicist who later turned into a historian of science. In 1965 he published in *Science* a paper on a pattern of bibliographic citations in science journal papers (Price 1965). It is probably the first example of a network exhibiting a scale-free power-law behavior. An explanation of this power law was later given by Price (1976). A citation network is very similar to a network of web pages. Price assumed that the rate at which a paper gets new citations by other papers is proportional to the number of citations this paper already received. The more highly cited a paper is already, the more likely it is that it will be cited by the next new paper; again the rich get richer. Price assumes that each paper can cite on average m other papers, with m fixed. This is approximately true because most journals have limits on the number of references a paper can have. In journals that have no limits, in practice the number of cited references in an average paper does not vary that much. Since each paper on average cites m other papers, the average number of citations a paper receives within this network is also m. Let N_k be the number of papers (nodes) in the network that have been cited k times, and let n be the total number of papers in the whole network. We write $p_k = N_k/n$ as the fraction of the papers in the network having k citations. As a new paper is published, the network size increases by 1, from n to $n + 1$. We can use n as a timelike quantity, although it increases by discrete integers only. So the change in N_k with the arrival of this new paper is given by

$$N_k(n + 1) - N_k(n) \equiv (n + 1)p_k(n + 1) - np_k(n).$$

This change is given by the number of papers that had originally been cited only $k - 1$ times and are now cited by this new paper, thus becoming papers with k citations. It is decreased by the number of papers that already had k citations and that this new citation moves them into the category of papers with $k + 1$ citations, therefore no longer belonging to the category of papers with k citations. The probability that a new paper will cite any paper that already has k citations is proportional to k, which is a central assumption of Price, as mentioned above. This assumption, however, has a slight technical difficulty: the new paper has no one citing it yet, and so $k = 0$. This assumption then ensures that no future papers will ever cite it. Price got around this difficulty by saying that one should count the citation in such a way that a new paper should come with one citation, that by itself. Then the probability that a new paper will cite any paper that has k citations is proportional to $(k + 1)$, and in fact equal to

$$\frac{(k+1)p_k}{\sum_k (k+1)p_k} = \frac{(k+1)p_k}{(m+1)},$$

since $\sum p_k = 1$ and $\sum k p_k = m$, the mean number of citations in the network. Furthermore, since the mean number of new citations by each new paper is m, the number of new citations of existing papers already with k citations is m times the above probability, and thus $(k+1)p_k m/(m+1)$. Therefore,

$$N_k(n+1) - N_k(n) = kp_{k-1}(n)\frac{m}{m+1} - (k+1)p_k(n)\frac{m}{m+1}. \qquad (2.13)$$

For large networks (n large), the fraction $p_k(n+1) \sim p_k(n)$. Thus $N_k(n+1) - N_k(n) \equiv (n+1)p_k(n+1) - np_k(n) \sim p_k(n)$. Equation (2.13) becomes

$$p_k = \frac{m[kp_{k-1} - (k+1)p_k]}{(m+1)}.$$

This is the same as

$$p_k = \frac{kp_{k-1}}{(k+2+1/m)}. \qquad (2.14)$$

This is a first-order difference equation, except that the coefficient is dependent on the variable (the index k). Nevertheless, one can simply iterate. Starting with $k = 1$, we get from Eq. (2.14) $p_1 = p_0/(3 + 1/m)$. With $k = 2$, it is $p_2 = 2p_1/(4 + 1/m) = 2p_0/(4 + 1/m)(3 + 1/m)$, and so on. It then yields the following general solution:

$$p_k = \frac{k(k-1)(k-2)\cdots(2)(1)}{(k+2+1/m)(k+1+1/m)\cdots(3+1/m)}p_0. \qquad (2.15)$$

It turns out that this unwieldy expression, when plotted, looks like a power law for large k:

$$p_k \propto k^{-(2+1/m)}. \tag{2.16}$$

There are two ways to demonstrate this. The long fraction in (2.15) is not just any fraction; it is actually a special function, Legendre's beta function, that mathematicians have studied. It is known that for large k it has the asymptotic behavior of (2.16). Another way to show this is to observe that, for large k, the difference equation (2.14) can be written approximately as

$$\frac{p_k}{p_{k-1}} \approx \left(1 - \frac{(2+1/m)}{k}\right). \tag{2.17}$$

This is because, using the binomial expansion of $1/(1+x) \approx 1-x$,

$$\frac{1}{(k+2+1/m)} = \frac{1}{k\left(1 + \frac{(2+1/m)}{k}\right)} \approx \frac{1}{k}\left(1 - \frac{(2+1/m)}{k}\right).$$

To solve Eq. (2.17), we use the trial solution

$$p_k = k^\lambda,$$

where the exponent λ is unknown and to be determined in the solution process. When substituted into (2.17), the trial solution yields

$$\frac{k^\lambda}{(k-1)^\lambda} \approx \left(1 + \frac{\lambda}{k}\right) = \left(1 - \frac{(2+1/m)}{k}\right),$$

from which we obtain the exponent to the power law:

$$\lambda = -(2+1/m).$$

Equation (2.16) states that the fraction of papers with k citations by other papers varies with k in a power law of the form of $\sim k^{-\gamma}$, with a negative exponent of $\gamma = 2 + 1/m$, slightly larger than 2. Price thought that this was consistent with what he found in the *Science Citation Index*.

We now return to the model of Barabási and Albert for the World Wide Web links, which is actually simpler than that of Price but is just a little more difficult to understand. In the simple model used by Barabási and Albert (1999), it is assumed that each new addition to the web initially contains m links, which are directed both ways. That is, it is assumed that this new site links to m other sites and that these other sites also have links to it. This is unrealistic but gets around the problem facing Price of the new additions having zero probability of being linked by other nodes. Because of these two-way links, the mean number of links in the network is now $2m$ instead of m in Price's model. With this simple modification the problem is solvable exactly. This is left as an exercise (exercise 7). It should be pointed out that Barabási and Albert (1999) were mainly interested in showing that the web links are "scale free," i.e., satisfy a power law. They were not too concerned with the exact magnitude of the exponent. Their result of an exponent $\gamma = 3$ was not too realistic. That problem was left to later authors. See Newman (2003).

2.10 Exercises

1. Aorta radius

Show that the radius of the aorta blood vessel scales with the mass M of the animal as $M^{3/8}$.

2. A critique

Two Polish scientists, Kozlowski and Konarzewski (2004), wrote a critique of the West, Brown, and Enquist (1997) paper, pointing out that the model is mathematically incorrect. Their criticism centers on the expression of Eq. (2.4) for an animal's body volume: $V = \pi \rho^2 l_N N_N \sim C l_N^3 N_N$. Since it is proportional to the mass of the animal, the authors pointed out that it then follows from Eq. (2.4) and the assumption that the properties of the capillaries are invariant with respect to an animal's size that the total number of capillaries N_N in an animal of mass M should scale linearly with M, and not as a $\frac{3}{4}$ power of M as in Eq. (2.9). Therefore, the final result, Kleiber's law of $\frac{3}{4}$ powers derived in Eq. (2.11), is wrong, according to these authors. In rebuttal, Brown, West, and Enquist (2005) wrote:

> WBE clearly state that only the characteristics of the capillaries themselves are assumed to be invariant. Nevertheless, K&K incorrectly interpreted this size-invariance to mean that each capillary must supply a constant volume of tissue. . . . Predicting the scaling of the "service volume" of tissue supplied by a capillary is an integral part of the WBE theory. WBE proved that the service volume increases with body size, as $M^{1/4}$.

Well, in the original paper, West, Brown, and Enquist (1997) assumed that the service volume surrounding each capillary of length l_N is a sphere of diameter l_N. This then leaves no room for the service volume to increase with body size. In our derivation, we have inserted a constant C for the service volume in (2.4) so that this volume is Cl_N^3. C is independent of k, the level of branching for an animal, but it may be different for different animals. Allowing C to depend on M, show that C is proportional to $M^{1/4}$, thus resolving the apparent inconsistency.

3. Self-similarity assumption

Etienne, Apol, and Olff (2006) gave a more general derivation of the $\frac{3}{4}$ power law that does not require the assumption of self-similarity. Read this article and rederive the $\frac{3}{4}$ power scaling for the metabolic rate result under this more general condition.

4. Universal law of growth

(*Note*: Solution of this problem requires a knowledge of how to solve a simple first-order ordinary differential equation. If you are unfamiliar with ordinary differential equations, please read the review in Appendix A first.) In a growing organism, metabolism supplies energy to both maintain existing tissues and create new tissues by cell division. Let Y be the metabolic rate of an organism, Y_c the metabolic rate of a single cell, $N_c(t)$ the total number of cells at time t, m_c the mass of a cell, and E_c the energy required to create a new cell. The cell properties, E_c, m_c, and Y_c, are assumed to be constant and invariant with respect to the size of the organism. Thus:

$$Y = Y_c N_c + E_c \frac{dN_c}{dt}.$$

Let m be the total body mass of the organism at time t, and $m = m_c N_c$. (Note that N_c is the total number of cells in a body and is proportional to mass m, while the total number of capillaries N_N is proportional to $\frac{3}{4}$ power of m.) From (2.11), we have $Y = Y_0(m)^{3/4}$.

a. Show that the above equation can be written as

$$\frac{dm}{dt} = am^{3/4} - bm,$$

with $a = Y_0 m_c / E_c$ and $b = Y_c / E_c$.

b. Let $m = M$ be the mass of a matured organism, when it stops growing (i.e., $dm/dt = 0$). Find M, and show that the above equation can be rewritten as

$$\frac{dm}{dt} = am^{3/4}[1 - (m/M)^{1/4}].$$

c. Let $r = (m/M)^{1/4}$, and $R = 1 - r$. Then the above equation becomes

$$\frac{dR}{dt} = -\left(\frac{a}{4M}\right)R.$$

Solve this simple ordinary differential equation and show that a plot of $\ln(R(t)/R(0))$ vs. the no-dimensional time $at/(4M^{1/4})$ should yield a straight line with a slope -1 for any organism regardless of its size.

d. Based on this scaling for time t, argue that, for a mammal, the interval between heartbeats should scale with its size as $M^{1/4}$.

5. Fractal dimensions

The intuitive concept of dimensions of objects can be stated in a mathematical form in the following manner: a volume V, which depends on lengths x, y, z, ... , becomes another volume V' when the lengths are multiplied by the same factor λ. Thus

$$V' = V(\lambda x, \lambda y, \lambda z, \ldots).$$

If, e.g., $\lambda = 2$, then V' will be $8 = 2^3$ times V when its length, width, and height are all increased by a factor of 2. This relationship is expressed in the following self-similar scaling relation:

$$V' = V(\lambda x, \lambda y, \lambda z, \ldots) = \lambda^d V(x, y, z, \ldots).$$

In this case $d = 3$, and that gives the dimension of the volume. Similarly, we can scale a surface area as

$$S' = S(\lambda x, \lambda y, \ldots) = \lambda^d S(x, y, \ldots)$$

and deduce that the dimension of the surface is $d = 2$. When defined in this more general way, the dimension of an object could give some nonintuitive values, including fractional numbers—hence the word fractal. For a visual example, see exercise 5 of Chapter 1.

Now consider the internal surface across which nutrients are exchanged in a biological organism. As an example, consider the lung, which has a compact and smooth outer surface (the chest wall), and an irregular internal surface enclosing each of a huge number of small air sacs. We are interested in determining the fractal dimension of the internal surface. The area, denoted by A, depends on the structure of the branching network characterized by length l_k at each level. So we write

$$A = A(l_0, l_1, l_2, \ldots).$$

Now we want to see how this area scales if all lengths are multiplied by the factor λ:

$$A' = A(\lambda l_0, \lambda l_1, \lambda l_2, \ldots) = \lambda^d A(l_0, l_1, l_2, \ldots),$$

and the dimension d should be 2 for a surface, as can be argued intuitively.

Now comes the twist. In the West, Brown, and Enquist model, all lengths l_k scale with size, except the terminal unit, which is denoted here by l_0. This terminal unit is invariant to the scaling. This is not the typical fractal self-similarity, and it is somewhat unfortunate that we need to introduce the fractal definition of dimensions while talking about a concept that is not fractal. The biologically relevant scaling needed is instead:

$$A'' = A(l_0, \lambda l_1, \lambda l_2, \ldots) = \lambda^{2+\epsilon} A(l_0, l_1, l_2, \ldots).$$

The exponent ϵ is no longer 0 because of this strange scaling, and is as yet undetermined. The above form of the scaling is only a hypothesis, because it has not been shown that such a surface can be scaled this way. Similarly, it is assumed that volume scales as

$$V'' = V(l_0, \lambda l_1, \lambda l_2, \ldots) = \lambda^{3+\sigma} V(l_0, l_1, l_2, \ldots).$$

The exponent σ is as yet undetermined.

a. Show that as far as the scaling with respect to λ is concerned, surface scales with volume as

$$A'' \propto V''^{\frac{2+\epsilon}{3+\sigma}}.$$

Argue that the superscript $''$ can be dropped, resulting in

$$A \propto V^{\frac{2+\epsilon}{3+\sigma}} \propto M^{\frac{2+\epsilon}{3+\sigma}}.$$

$\epsilon = \sigma = 0$ yields the $\frac{2}{3}$ power law of Rubner, while $\epsilon = \sigma = 1$ would give the $\frac{3}{4}$ power law of Kleiber.

b. West, Brown, and Enquist (1999) suggested that organisms have evolved to maximize their internal surface within a given volume to facilitate exchange of nutrients, and they claim that this is equivalent to maximizing the exponent $b = (2 + \epsilon)/(3 + \sigma)$. This expression does not have a distinct maximum unless one allows negative numbers for σ. So, clearly, more assumptions are needed.

Use the following arguments to show that the maximum is attained for $\epsilon = 1$ and $\sigma = \epsilon$: a volume cannot have a dimension less than a surface; the smallest value that σ can take is ϵ. On the other hand, the dimension of the internal surface, being an area, cannot exceed the dimension of an ordinary volume; the largest value ϵ can take is 1. Hence derive Kleiber's $\frac{3}{4}$ power law.

c. Summarize, in your own words, the assumptions needed to derive Kleiber's law vs. those needed for Rubner's law. Are you satisfied with these assumptions in this fractal context?

6. *Zipf's law for random texts*

Following Li (1992), we let there be M letters and one space that our random monkey can type, each with probability $p = 1/(M + 1)$. A word is formed when a string of letters of any size is preceded and followed by a space, denoted by _ here. The probability of getting the word _a_ is p^3. The probability of finding a three-letter word, such as _bst_, is p^5, and that of an L-letter word, p^{L+2}. So the frequency of occurrence of any word with length L is

$$f_i = cp^{L+2},$$

where c is a normalization constant. There are M^L such words with length L. The constant c is to be determined by the convention that the frequency of occurrence of all words is normalized to 1:

$$1 = \sum_{L=1}^{\infty} M^L \frac{c}{(M+1)^{L+2}}.$$

The frequency of occurrence of all words of length L is

$$f(L) = M^L f_i(L).$$

a. Show that

$$f(L) = \frac{M^{L-1}}{(M+1)^L}.$$

b. To convert word length L into rank, we note that the formula for f_i implies that shorter words (those with smaller values of L) should occur more frequently and hence have a lower value for the rank $k(L)$. The rank is easy to understand (the most frequently occurring word is of rank 1, the next is rank 2, etc.) but it is difficult to define mathematically. Nevertheless, argue that the rank can be expressed as

$$k(L) = \alpha M^L - \beta,$$

for some constants α and β. This yields, upon taking the log with base M:

$$L = \log_M(\alpha^{-1}(k + \beta)).$$

Substituting this expression for L into the formula in (a), show that f satisfies the following generalized Zipf's law:

$$f = C(k + \beta)^{-b},$$

where

$$b = \frac{\ln(M + 1)}{\ln(M)}.$$

Show that $b \sim 1$.

7. A mathematical model for the World Wide Web

The model of Barabási and Albert considers the situation when a new node attaches to the existing network consisting of n nodes. This new node has m undirected links, meaning that it is linked to m existing nodes in two directions. If the entire web is built up this way, then the mean number of links is $2m$. The probability of attachment by this new node to any existing node that already has k links is

$$\frac{k p_k(n)}{2m},$$

where $p_k(n)$ is the fraction of nodes in the network of n nodes that have k links. So the number of nodes that gain one link by this additional node is m times this probability.

a. Derive the counterpart of Eq. (2.13) by arguing that its right-hand side should be

$$\frac{1}{2}(k - 1)p_{k-1}(n) - \frac{1}{2}kp_k(n).$$

b. For a large network (n large), show that this reduces to

$$\frac{p_k}{p_{k-1}} = \frac{k-1}{k+2}.$$

c. Solve the first-order difference equation above by iteration and show that the solution is a constant times

$$p_k \propto \frac{1}{(k+2)(k+1)(k)}.$$

Therefore, for large k it is a power law with $\gamma = 3$.

3

Modeling Change One Step at a Time

Mathematics required:
> derivatives as limits to differences

Mathematics developed:
> solution methods for linear first-order ordinary differential
> equations; solution to difference equations through iteration

3.1 Introduction

Let us do something simple but practical for a change: compound interest and mortgage payments. Without a strategy, one can get overwhelmed by the complexity very quickly even for this simple problem. We will introduce the *difference equation* as a modeling tool. It allows us to consider changes in time from one step to another, *one step at a time.*

3.2 Compound Interest and Mortgage Payments

Consider three problems of increasing complexity, all involving compound interest.

Your Bank Account

Let $P(t)$ be your account balance at time t. $P(0) = P_0$ is your initial deposit. From then on, it is earning interest at a fixed *interest rate* r. The quantity r has the dimension of $(\text{time})^{-1}$. For example, $r = 6\%$ per year. Without compounding, the interest you earn in one year is simply $(r P_0 \cdot 1\text{yr})$ and so

$$P(1\text{yr}) = P_0(1 + r \cdot 1\text{yr}).$$

Often bank savings and certificates of deposit carry compound interest. Let Δt be the *compounding interval*. If your balance is compounded monthly, then $\Delta t = 1\text{month} = \frac{1}{12}\text{yr}$. One month after your initial deposit, you earn interest of

$$P_0 \cdot r \cdot \Delta t.$$

That is added to your account, which becomes

$$P_0 \cdot (1 + r \Delta t).$$

This larger amount then earns interest for the next month at the rate r. At the end of two months, your balance becomes

$$P(2\Delta t) = P(\Delta t) \cdot (1 + r \Delta t),$$

and so on.

Since there is no interest compounding within Δt, the equation governing the change of your balance in one Δt step is actually quite simple:

$$P(t + \Delta t) = P(t) \cdot (1 + r \Delta t). \tag{3.1}$$

This is a *difference equation*. It describes how $P(t)$ evolves from one time step to another. We have accomplished our modeling process for this problem of compound interest once we have written down Eq. (3.1). Solving Eq. (3.1) is simply a mathematical exercise.

Solution
Equation (3.1) can be solved either by assuming $P(t) = c\lambda^t$ as before, or by iteration, starting with $t = 0$:

$$P(\Delta t) = P_0 \cdot (1 + r \Delta t),$$

$$P(2\Delta t) = P(\Delta t) \cdot (1 + r \Delta t) = P_0 \cdot (1 + r \Delta t)^2,$$

$$\vdots$$

$$P(m\Delta t) = P_0 \cdot (1 + r \Delta t)^m.$$

This last equation can be rewritten as

$$P(t) = P(0) \cdot (1 + r \Delta t)^{t/\Delta t}. \tag{3.2}$$

This yields the balance at time t when the interest rate is r and the compounding period is Δt.

Your Mortgage Payments, Monthly Interest Compounding

Suppose you borrow P_0 from the bank to buy a house and agree to pay interest at the fixed rate r compounded monthly. You also agree to pay the bank a fixed amount M monthly. Suppose the loan is for 30 years. After 360 monthly payments you pay off both the original principal and interest completely. What is M, and how much total interest have you paid when it is all over?

The problem is simpler if we consider the change over a single Δt. In one Δt, the principal increases because of the simple interest, but reduces by the monthly payment of M. Thus

$$P(t + \Delta t) = P(t) \cdot (1 + r \Delta t) - M. \tag{3.3}$$

After we solve Eq. (3.3), we set $P(0) = P_0$ and $P(30\,\text{years}) = 0$, and we will get M.

Solution

Let $R = (1 + r \Delta t)$ be the growth factor due to interest accrued during Δt, just before a payment is made. Equation (3.3) is, in terms of R,

$$P(t + \Delta t) = P(t) \cdot R - M.$$

It is solved by iteration:

$$P(\Delta t) = P_0 \cdot R - M,$$

$$P(2\Delta t) = P(\Delta t) \cdot R - M = [P_0 \cdot R - M] \cdot R - M,$$

$$P(3\Delta t) = P(2\Delta t) \cdot R - M = P_0 \cdot R^3 - M \cdot [1 + R + R^2],$$

$$\vdots$$

$$P(m\Delta t) = P_0 \cdot R^m - M \cdot [1 + R + R^2 + \cdots + R^{m-1}].$$

The sum inside the square brackets is a geometric series, which can be summed exactly using the following trick. Let

$$S \equiv 1 + R + R^2 + \cdots + R^{m-1},$$

$$RS = R + R^2 + \cdots + R^{m-1} + R^m,$$

$$S - RS = 1 - R^m.$$

Solving for S, we get

$$S = (1 - R^m)/(1 - R).$$

Therefore,

$$P(m\Delta t) = P_0 \cdot R^m - M\frac{1 - R^m}{1 - R}.$$

(3.4)

In a 30-year mortgage, we want

$$P(360\,\text{months}) = 0,$$

$$0 = P_0 \cdot R^{360} - M\frac{1 - R^{360}}{1 - R},$$

and we can solve for the correct monthly payment that will allow you to pay off your loan in 30 years:

$$M = P_0 \cdot R^{360} \cdot \frac{(R - 1)}{(R^{360} - 1)}.$$

(3.5)

Your Mortgage Payments, Daily Interest Compounding

Most mortgages carry a daily compounding interest. The repayment is still on a monthly basis. This introduces a complication because the interval over which interest compounds is different from the payment interval. We choose Δt to be the interval between mortgage payments. We will have, one Δt later,

$$P(t + \Delta t) = P(t) \cdot R - M.$$

However, the growth factor R is now due to interest compounded daily over a period of one month. For now we take one year to be 365 days and one month to be 365 days/12. Thus, from Eq. (3.2), for compounding interest without repayment, we have

$$R = \left(1 + r \cdot \frac{1}{365}\text{yr}\right)^{365/12}.$$

(3.6)

The solution is still Eq. (3.4) and Eq. (3.5), but with R replaced by Eq. (3.6).

3.3 Some Examples

Example 1

Suppose you borrowed \$100,000 to buy a condominium at 10% annual interest, compounded monthly. What should your monthly payment be if you want to pay off the loan in 30 years?

Solution

We want to find M such that $P(360\,\text{months}) = 0$. This yields from Eq. (3.5):

$$M = P_0 \cdot R^{360} \cdot \frac{(R-1)}{(R^{360} - 1)}$$

$$= \$100{,}000 \cdot \left(1 + \frac{0.1}{12}\right)^{360} \cdot \left(\frac{0.1}{12}\right) \bigg/ \left[\left(1 + \frac{0.1}{12}\right)^{360} - 1\right]$$

$$= \$877.57.$$

This is the required monthly payment if the \$100,000 original loan is to be paid off in 360 equal monthly installments.

Example 2

What is your monthly mortgage payment if you borrow \$100,000 for 30 years at 10% annual interest, compounded daily?

Solution

We use the same formula as in Example 1, except with

$$R = \left(1 + \frac{0.1}{365}\right)^{365/12} = 1.008367.$$

So

$$M = \$100{,}000 \cdot (1.00836)^{360} \cdot (0.00836)/[(1.00836)^{360} - 1]$$

$$= \$880.55.$$

The required monthly mortgage payment is \$881. You pay about \$3 more per month with daily compounding of the money you owe compared with monthly compounding.

3.4 Compounding Continuously

We would like to calculate the account balance for various compounding frequencies n within a year. We have found, from (3.2), that the balance after one year is

$$P(1\text{yr}) = P_0 \cdot \left(1 + r \cdot \left(\frac{1}{n}\text{yr}\right)\right)^n,$$

where n is the number of times within a year that interest is deposited into the account for the purpose of subsequent compounding. Table 3.1 lists the year-end balance for an initial $1,000 deposit under various compounding frequencies.

Continuous Compounding

If the interest is compounded continuously, $n \to \infty$. We have

$$P(1\text{yr}) = P_0 \cdot \lim_{n \to \infty} \left(1 + r \cdot \left(\frac{1}{n}\text{yr}\right)\right)^n.$$

The symbol e, in honor of the Swiss mathematician Leonhard Euler (1707–1783), is assigned to the limit:

$$\lim_{m \to \infty} \left(1 + \frac{1}{m}\right)^m = e = 2.71828\ldots.$$

By letting $\frac{1}{m} = r \cdot (\frac{1}{n}\text{yr})$, we can show that

$$\lim_{n \to \infty} \left(1 + r \cdot \left(\frac{1}{n}\text{yr}\right)\right)^n = \lim_{m \to \infty} \left(1 + \frac{1}{m}\right)^{m(r \cdot 1\text{yr})} = \left[\lim_{m \to \infty} \left(1 + \frac{1}{m}\right)^m\right]^{(r \cdot 1\text{yr})}$$

$$= e^{(r \cdot 1\text{yr})}.$$

So the formula for the balance after one year of continuous compounding is

$$P(1\text{yr}) = P_0 e^{r \cdot (1\text{yr})}.$$

It can be shown easily that at any time t,

$$\boxed{P(t) = P(0)e^{rt}.} \tag{3.7}$$

It is even simpler than the "discrete compounding" formula we wrote down earlier, as it only depends on r (no n-dependence) (see Table 3.1).

Table 3.1

Compounding interest at $r = 6\%$ on an original principal $P_0 = \$1,000$

Compounding Frequency	n	Year-End Balance	Annual Yield
Annually	1	$1,060.00	6%
Semiannually	2	$1,060.90	6.090%
Quarterly	4	$1,061.36	6.136%
Monthly	12	$1,061.68	6.168%
Daily	365	$1,061.83	6.183%
Instantly	∞	$1,061.84	6.184%

The "annual yield" is calculated from $(P - P_0)/P_0$ and is listed in the right column.

There is actually very little difference (0.016%) between the interest earned from continuous compounding and that from daily compounding. It amounts to about 1¢ on a principal of $1,000.

Double My Money: "Rule of 72," or Is It "Rule of 69"?

In financial circles there is a mythical "rule of 72." It says that if you divide the APR, the annual percentage rate (i.e., 100 times r times 1 year), into 72, you will get the number of years it takes to double your money. As is often the case with these "rules" created by nonscientists, there is no specification of the conditions under which the formula is valid. The accuracy of the rule actually depends on the magnitude of r and how it is compounded, as we will see. In any case, the rule is supposed to work this way: if you are earning 6% interest annually, APR = 6. Divide 6 into 72, and you get 12. So it takes 12 years to double your money at a 6% interest rate. Let us see if it is right.

First consider the case where the annual interest rate r is compounded daily (this is the case in most bank savings accounts). Since we have shown that there is very little difference in the yield between daily compounding and continuous compounding, we will use the latter to approximate the former. Starting with a principal of $P(0)$, after t years we have in our savings account

$$P(t) = P(0)e^{rt}.$$

We want to find the $t = \tau$ for which $P(\tau)/P(0) = 2$. Thus

$$2 = e^{r\tau}.$$

Take the natural log on both sides, and since $\ln 2 = 0.693$, we have

$$r\tau = \ln 2 = 0.693,$$

yielding

$$(\tau/\text{yr}) = 69/\text{APR}.$$

It seems that the rule should have been called the "rule of 69," and that it takes 11.5 years to double your money at an APR of 6.

Well, perhaps the financial people who came up with this rule of 72 were not thinking of daily or continuous compounding. Perhaps there was no compounding within each year. To resolve this ambiguity, let's use, instead of APR, the AY, the annual yield (in percent). So, regardless of how often we compound within a year, we always have

$$P(t) = P(0) \left(1 + \frac{AY}{100}\right)^{(t/\text{yr})}.$$

Again we seek the $t = \tau$ for which $P(\tau)/P(0) = 2$:

$$\ln 2 = (\tau/\text{yr})\ln\left(1 + \frac{AY}{100}\right),$$

so

$$(\tau/\text{yr}) = 0.693/\ln\left(1 + \frac{AY}{100}\right).$$

This formula is not so easy to use. We should not forget that most people on Wall Street may not know what ln is. Since an approximate formula will suffice, and we know that $\ln(1 + x) \cong x$ for $|x| \ll 1$, we approximate the above formula by

$$\tau \cong 69/AY \text{ years, provided that } |AY/100| \ll 1.$$

Thus we again obtained the rule of 69, for small annual yields.

Table 3.2 compares various approximations with the exact value of τ, the number of years it takes to double the original investment. Neither approximation is perfect, but the rule of 72 appears to give a better approximation to the exact value for AY between 6 and 8, a more commonly encountered range for interest rates, while the rule of 69 gives a better approximation for AY of 2 or less.

We thus see that the rule of 72 is an ad hoc formula because it is not derived mathematically. The rule of 69, on the other hand, is a mathematically correct (asymptotic) approximation for small AY; the approximation is better the smaller the AY. The rule of 69 is also almost exact for any interest rate as long as the compounding frequency is daily or more often.

TABLE 3.2
Years to double an original investment for various annual yields

AY	Exact	72/AY	69/AY
2	35.0	36	34.5
4	17.7	18	17.3
6	11.9	12	11.5
8	9.0	9.0	8.6
10	7.3	7.2	6.9
12	6.1	6.0	5.8
14	5.3	5.1	4.9
16	4.7	4.5	4.3
18	4.2	4.0	3.8
20	3.8	3.6	3.5
25	3.1	2.9	2.8
30	2.6	2.4	2.3
35	2.3	2.1	2.0
40	2.1	1.8	1.7
45	1.9	1.6	1.5
50	1.7	1.4	1.4
60	1.5	1.2	1.2
70	1.3	1.0	1.0
80	1.2	0.9	0.9
90	1.1	0.8	0.8
100	1.0	0.7	0.7

3.5 Rate of Change

The difference equation (3.1) for compound interest can also be written in the form of a rate of change:

$$\frac{P(t + \Delta t) - P(t)}{\Delta t} = r P(t). \qquad (3.8)$$

Equation (3.8) says that the time rate of change of $P(t)$ is proportional to $P(t)$ itself, and the proportionality constant is r. If the rate of change of $P(t)$ is in addition reduced by withdrawal at the *rate* of $w(t) \equiv W(t)/\Delta t$, we say

$$\frac{P(t + \Delta t) - P(t)}{\Delta t} = r P(t) - w(t). \qquad (3.9)$$

Equation (3.9) is the same as Eq. (3.3); it is just rewritten in a form that emphasizes that it is the rate of change that is being modeled.

Continuous Change

It is now quite easy to go to the continuous limit. We simply take the limit as $\Delta t \to 0$ and recognize that

$$\lim_{\Delta t \to 0} \frac{P(t + \Delta t) - P(t)}{\Delta t} = \frac{d}{dt} P(t).$$

Equation (3.8) reduces to the *differential* equation:

$$\frac{d}{dt} P(t) = r P(t), \tag{3.10}$$

whose solution is (see Appendix A)

$$P(t) = P_0 e^{rt},$$

the same as Eq. (3.7), obtained earlier.

Alternatively, one could view the solution to Eq. (3.10) as the continuous limit of Eq. (3.2) as $\Delta t \to 0$. The limiting process done to arrive at (3.7) then tells us that $P(t) = P_0 e^{rt}$ is the solution to Eq. (3.10) because that equation is the limit of the discrete equation (3.8) whose solution is (3.2). Equation (3.9) becomes, in the continuous limit:

$$\frac{d}{dt} P(t) = r P(t) - w(t).$$

Similar equations are used to model a variety of phenomena. One example that we will discuss later is the harvesting of fish (where $w(t)$ is the harvesting rate and $P(t)$ is the fish population), although the way the fish population "compounds" needs some modification.

3.6 Chaotic Bank Balances

Although we will wait until chapter 7 to discuss the phenomenon of *deterministic chaos*, we cannot resist the temptation to give one example here. Normally we expect our bank account balances to be precisely predictable (to the cent). Here is an example where a precise formula for calculating balances can yield chaotic values.

Most banks do not pay interest on "dormant accounts." In the state of Washington, your bank account is considered "dormant" if there are no deposits or withdrawals for 5 years. One possible reason for this practice is to avoid the astronomical sums that can build up through interest compounding over a long period of time, which can bankrupt the bank if later claimed.

An alternative way to avoid this is to specify a maximum amount, K, that a bank is willing to pay out to any account, and have the interest rate reduced gradually as the account balance approaches K. For continuously compounded interest, a suitable formula for the account balance $P(t)$ at an initial interest rate r is

$$\frac{d}{dt}P(t) = r \cdot P(t) \cdot \left(1 - \frac{P(t)}{K}\right).$$

The depositor and the bank can agree on this formula beforehand. The balance can then be calculated automatically without human intervention.

Now, since most banks do not pay interest continuously, a discrete version is more appropriate. We consider the following, rather reasonable (it seems), discrete formula:

$$\frac{P(t + \Delta t) - P(t)}{\Delta t} = r \cdot P(t) \cdot \left(1 - \frac{P(t)}{K}\right),$$

where Δt is the compounding period, taken to be one year for our present problem. It will also be convenient to normalize all $P(t)$ by K. For example, if $K = \$1$ million, then all dollar amounts will be measured in millions of dollars. So, for $p(t) \equiv P(t)/K$, the formula can be rewritten as

$$p(t + \Delta t) = p(t) + r\Delta t \cdot p(t) \cdot (1 - p(t)).$$

The bank pays a huge interest of 300% per year compounded yearly (i.e., $r\Delta t = 3$). Sensing an opportunity, you sell your house and put the proceeds, \$58,000 ($p(0) = 0.058$), in the bank.

    a. Calculate your account balance after 45 years. Since most banking institutions want to be accurate to the penny, keep nine decimal places in your calculator for $p(t)$.

    b. You want to be more accurate and keep an extra digit. Do the calculation in (a) again, but this time keep 10 decimal places in your calculator.

    c. Suppose you want to withdraw your money after 43 years. If you had a choice, would you prefer to have the more "accurate" way of computing interest (i.e., keep 10 decimal places instead of 9)?

Solution

Table 3.3 lists the values of $P_n \equiv p(n\Delta t)$ for Calculator 1, which keeps 9 digits after the decimal point for $P(t)$ (and hence is accurate to the cent), and for Calculator 2, which keeps 10 digits (and hence is accurate to 0.1 cent). $n = 1$ is one year later, $n = 2$ is two years later, etc.

TABLE 3.3
Balance P_n after n years

n	P_n (Calculator 1)	P_n (Calculator 2)	n	P_n (Calculator 1)	P_n (Calculator 2)
0	$58,000.000	$58,000.0000	23	$780,938.022	$780,094.1514
1	$221,908.000	$221,908.0000	24	$1,294,159.505	$1,294,735.9504
2	$739,902.518	$739,902.5186	25	$152,091.546	$149,920.2578
3	$1,317,242.863	$1,317,242.8633	26	$538,970.668	$532,252.7801
4	$63,585.171	$63,585.1704	27	$1,284,414.529	$1,279,132.0546
5	$242,211.462	$242,211.4599	28	$188,496.069	$207,991.7790
6	$792,846.671	$792,846.6656	29	$647,391.971	$702,185.3756
7	$1,285,569.152	$1,285,569.1569	30	$1,332,218.791	$1,329,548.5972
8	$184,212.474	$184,212.4560	31	$4,454.442	$15,095.9718
9	$635,047.189	$635,047.1371	32	$17,758.241	$59,700.2221
10	$1,330,333.959	$1,330,333.9493	33	$70,086.898	$228,108.5388
11	$11,970.508	$11,970.5472	34	$265,611.072	$756,333.6387
12	$47,452.152	$47,452.3067	35	$850,796.563	$1,309,212.8357
13	$183,053.487	$183,054.0625	36	$1,231,621.877	$94,736.5953
14	$631,688.210	$631,689.8806	37	$375,810.164	$352,021.3137
15	$1,329,662.856	$1,329,663.2066	38	$1,079,540.817	$1,036,328.2389
16	$14,641.492	$14,640.0974	39	$821,938.141	$923,384.2993
17	$57,922.848	$57,917.3922	40	$1,261,005.641	$1,135,621.5046
18	$221,626.223	$221,606.2958	41	$273,616.884	$673,577.4132
19	$739,150.343	$739,097.1321	42	$869,868.938	$1,333,190.0580
20	$1,317,571.683	$1,317,594.8163	43	$1,209,459.844	$573.0397
21	$62,301.312	$62,210.9653	44	$449,460.003	$2,291.1736
22	$237,560.887	$237,233.2485	45	$1,191,797.168	$9,148.9459

Calculator 1 is accurate to the cent, while Calculator 2 is accurate to 0.1 cent.

a. An initial deposit of $58,000 grows to $1.19 million after 45 years.
b. It is, however, only $9,149 after 45 years if we keep 10 digits after the decimal point.
c. After 43 years, I would prefer to keep the $1.209 million using the bank's method. My "more accurate" method gives me only $573, a tenth of my original deposit!

3.7 Exercises

1. Mortgage

a. You borrowed $200,000 on a 15-year mortgage at a 4.75% annual interest rate compounded daily. What is your monthly mortgage payment?

b. Same as in (a), except that it is a 30-year mortgage at 5.25%. What is your monthly payment? (Typically, the interest rate for a 30-year mortgage is higher than for a 15-year mortgage by about 0.50%.)

c. How much more total interest will you pay on a 30-year mortgage compared with a 15-year mortgage ((b) vs. (a))?

2. Biweekly mortgage

Let M be the monthly payment for a 30-year mortgage at r annual interest rate compounded daily. Your employer pays you on a biweekly schedule instead of on a monthly schedule. You want to make a mortgage payment every two weeks in the amount of $M/2$. How many years sooner can you finish paying off your mortgage? Use $r = 10\%$ per year, $P_0 = \$100,000$. Your answer should be independent of P_0.

3. Redo problem 1, but use continuously compounding interest as an approximation to daily compounding. What errors would you have in your answers for (a), (b), and (c)?

4. Lottery winner

You are the winner of the $10 million lottery jackpot. The first decision you need to make is whether to take your winnings in 25 annual payments of $400,000 each, or to elect the $10 million lump sum up front. Discuss how you arrive at your decision. Your decision should be dependent on the prevailing interest rate for safe investments.

5. Power of compounding

A professor's daughter is now at a private college that costs $30,000 per year. When she was born her grandparents put $10,000 in a college fund for her, investing it in a mutual fund that has had an average annual return of 18% for the past 18 years. Ignore the year-to-year fluctuations of the return. Does she have enough money in her account for four years of college expenses if she entered college at 18 years of age?

6. Power of compounding

a. You borrowed $1,000 from a loan shark at 5% *monthly* interest. How much do you owe four years later?

b. You are trying to build a nest egg for your retirement. You have estimated that you will need income of $5,000 a month in order to live comfortably after retirement. How large a nest egg (principal P_0) must you have at retirement? Assume that at your retirement you put that money, P_0, in an annuity with a guaranteed annual return of 8%. And suppose you think you will live forever.

c. You are now 25 and plan to retire at age 65. You want to start saving so that you can build a nest egg of $1 million. How much should you save each month? Assume that your savings will be earning 10% interest each year, compounded monthly.

7. *Present value of money*

You have a contract that entitles you to receive $1 million 20 years from now. But you can't wait and want your money now. You want to sell your contract. What is a fair price for it? Assume the risk-free, inflation-adjusted interest rate is 3% per year, compounded continuously.

4

Differential Equation Models: Carbon Dating, Age of the Universe, HIV Modeling

Mathematics required:
> solution of first-order linear ordinary differential equations as reviewed in Appendix A

4.1 Introduction

Towards the end of the last chapter, we showed that in the limit of Δt approaching 0, difference equations become differential equations. Differential equations are good approximations to situations where there are a large number of events happening on average and when the time scale over which we are examining these averages is much longer than the interval between events.

Decay of radioactive elements is often modeled using differential equations. In the 1940s an American chemist, W. F. Libby, developed the technique for carbon dating; for this work he received the Nobel Prize in Chemistry in 1960. In a recent letter to *Nature*, Cayrel et al. (2001) reported the most accurate determination of the age of our universe so far, using uranium line emission from a very old, metal-poor star—known as CS31082-001—near the edge of our Milky Way.

During the early days of AIDS research, it was not understood that in HIV-infected individuals the virus was being produced at a prolific rate even before the onset of symptoms and full-blown AIDS. The modeling effort of Perelson et al. (1996) reported in *Science* quantified this viral production rate and did much to advance the treatment strategies of David Ho (1996 *Time* Person of the Year) and others.

4.2 Radiometric Dating

Radiometric dating methods are based on the phenomenon of radioactivity, discovered and developed at the beginning of the 20th century

TABLE 4.1
Some radioactive isotopes and their half-lives

Substance	τ, half-life
Xenon-133	5 days
Barium-140	13 days
Lead-210	22 years
Strontium-90	25 years
Carbon-14	5,568 years
Plutonium	23,103 years
Uranium-235	0.707×10^9 years
Uranium-238	4.5×10^9 years

by the British physicist Lord Ernest Rutherford (1871–1937) and others. Certain atoms are inherently unstable, so that after some time and without any outside influence they will undergo transition to an atom of a different element, emitting in the process radiation (in the form of alpha particles or beta particles, which are detected by a Geiger counter). From experimental evidence, Rutherford found that the rate of decay is proportional to the number of radioactive atoms present in the sample,

$$\frac{d}{dt}N = -\lambda N, \tag{4.1}$$

where N is the number of atoms in a radioactive sample at time t and λ is a positive constant, known as the *decay constant*. This constant has different values for different substances.

Experimentally measured values for the *half-life*, τ, are used to find λ. τ is defined as the time taken for half of a given quantity of atoms to decay (Table 4.1). So if $N = N_0$ at $t = 0$, then $N(\tau) = N_0/2$.

Since the solution to Eq. (4.1) is

$$N(t) = N_0 e^{-\lambda t}, \tag{4.2}$$

$$N(\tau)/N_0 = e^{-\lambda \tau} = \frac{1}{2},$$

then

$$\lambda = \frac{\ln 2}{\tau}. \tag{4.3}$$

(So for carbon-14, an isotope of carbon, $\lambda = 1.245 \times 10^{-4}$ per year.)

4.3 The Age of Uranium in Our Solar System

The heavy elements we find on earth were not produced here. Some (up to iron) were produced in the nuclear furnace of the sun. Elements heavier than iron were produced elsewhere, most likely in a supernova, whose explosion spewed elements such as uranium into the "stardust." The solar system then formed out of these elements. So the uranium we now find on earth probably originated in a recent supernova explosion nearby. Nuclearsynthesis theory, which we will not go into here, tells us that the heavy elements close to each other in atomic number should have been produced in nearly equal concentrations.

The two isotopes of uranium closest to each other found on earth are ^{238}U and ^{235}U. The superscripts indicate the atomic number: the sum of the number of protons and the number of neutrons in the element. Both elements are radioactive. ^{235}U decays faster, with a half-life of 0.707Gyr, while ^{238}U's half-life is 4.468Gyr (1Gyr $\equiv 10^9$ years). The *isotopic ratio* ^{238}U$/^{235}$U is measured at present to be 137.8. We do not know each isotope's original abundance.

If we find the age of the uranium (and hence the time of the nearby supernova explosion), we will obtain an upper bound on the age of our solar system.

The concentration of each element, $U_1(t) \equiv {}^{238}U(t)$ and $U_2(t) \equiv {}^{235}U(t)$, decays according to Eq. (4.1),

$$\frac{d}{dt}U_1 = -\lambda_1 U_1,$$

$$\frac{d}{dt}U_2 = -\lambda_2 U_2,$$

where $\lambda_1 = \ln2/4.47\text{Gyr}$, $\lambda_2 = \ln2/0.707$ Gyr.

Solving, we find

$$U_1(t) = U_1(0)e^{-\lambda_1 t}, \quad U_2(t) = U_2(0)e^{-\lambda_2 t}.$$

Unfortunately we do not know the initial isotopic concentrations: $U_1(0)$ and $U_2(0)$. However, we have some theoretical idea about their isotopic ratio. Therefore, we form the ratio

$$U_1(t)/U_2(t) = (U_1(0)/U_2(0))e^{-(\lambda_1 - \lambda_2)t}. \tag{4.4}$$

From nuclearsynthesis theory, we have

$$U_1(0)/U_2(0) \cong 1.$$

From measurement at the present time t, we have

$$U_1(t)/U_2(t) = 137.8.$$

Substituting these numbers into (3.4), we get

$$137.8 = 1 \cdot e^{(\lambda_2 - \lambda_1)t}.$$

Thus

$$t = \frac{\ln(137.8)}{\lambda_2 - \lambda_1} \cong 5.97 \text{Gyr}.$$

Our solar system is at most six billion years old.

4.4 The Age of the Universe

In a recent letter to *Nature*, Cayrel et al. (2001) reported the most accurate determination of the minimum age of the universe so far. A headline in the *PhysicsWeb* of February 7, 2001, reads:

> **Uranium Reveals the Age of the Universe**
> Astronomers have spotted for the first time the fingerprint of uranium-238 in an ancient star—and have used it to make the most reliable guess yet of the age of the universe. Roger Cayrel of the Observatoire de Paris-Meudon, France, and colleagues have used a kind of "stellar carbon dating" to estimate the age of the star—and therefore the minimum age of the universe. The new estimate makes the universe 12.5 billion years old—give or take 3 billion years. (R. Cayrel et al., 2001, *Nature*, **409**, 691)

An excerpt from the Cayrel et al. paper describes the rationale and the method:

> The ages of the oldest stars in the Galaxy indicate when star formation began, and provide a minimum age for the universe. Radioactive dating of meteoritic material and stars relies on comparing the present abundance ratios of radioactive and stable nuclear species to the theoretically predicted ratio of their production. The radio isotope ^{232}Th (half-life 14.05Gyr) has been used to date Galactic stars, but it decays by only a factor of two over the lifetime of the universe. ^{238}U (half-life 4.468Gyr) is in principle a more precise age indicator, but even its strongest spectral line, from singly ionized uranium at a wavelength of 385.957 nm, has previously not been detected in stars.

Figure 4.1. The ancient star CS31082-001
surrounded by the Milky Way star field.
(Courtesy of ESO.)

This is because other metals, such as iron, have spectral lines in nearby wavelengths and often obscure the emission line of uranium. Fortunately, one very old star—CS31082-001—near the edge of our Milky Way was found by the authors to be very metal-poor. (See Figure 4.1.) Metals were scarce in very old stars because very few super-novae, which create metals, had yet exploded. The traces of uranium-238 in the star's atmosphere could have come from just one supernova in the early history of the universe.

Ideally, one should use the abundance ratio of the two closest iso-topes of uranium, $^{238}U/^{235}U$, to determine the age of uranium in that old star. No ^{235}U emission was found. This is understandable: because of its rather short half-life (the time for it to decay to half its original num-ber of atoms is 0.707Gyr), there is probably not much ^{235}U left by now. The next choice is to use the stable elements osmium (Os) and iridium (Ir), whose emission lines have also been detected from this old star.

Since absolute abundances cannot be measured, abundance ratios are often used. The ratios $^{238}U/Os$ and $^{238}U/Ir$ are measured. Os and Ir are stable. The presently measured (at time t) ratios are

$$\log_{10}(^{238}U/Os) = -2.19 \pm 0.18,$$
$$\log_{10}(^{238}U/Ir) = -2.10 \pm 0.17.$$

The ratios, when produced (at time 0), are not 1 because the elements are not close to each other in atomic number. Nuclearsynthesis models predict the following theoretical values at $t = 0$:

$$\log_{10}(^{238}\text{U/Os})_0 = -1.27,$$
$$\log_{10}(^{238}\text{U/Ir})_0 = -1.30.$$

Let $U(t)$ denote the concentration of ^{238}U at the present time, and $t = 0$ is the time since it was created. It decays according to

$$U(t) = U(0)e^{-\lambda_1 t}, \quad \text{where } \lambda_1 = \frac{\ln 2}{4.47\text{Gyr}}.$$

For stable elements,

$$Os(t) = Os(0),$$
$$Ir(t) = Ir(0).$$

So,

$$(U(t)/Os(t)) = (U(0)/Os(0))e^{-\lambda_1 t},$$
$$\ln(U(t)/Os(t)) = \ln(U(0)/Os(0)) - \lambda_1 t.$$

Similarly

$$\ln(U(t)/Ir(t)) = \ln(U(0)/Ir(0)) - \lambda_1 t.$$

Therefore

$$t = 6.45\text{Gyr}[\ln(U/Os)_0 - \ln(U(t)/Os(t))]$$
$$= 14.8\text{Gyr}[\log_{10}(U/Os)_0 - \log_{10}(U(t)/Os(t))]$$
$$= 14.8\text{Gyr}[-1.27 + 2.19]$$
$$= 13.6\text{Gyr}.$$

Similarly, we have

$$t = 14.8\text{Gyr}[\log_{10}(U/Ir)_0 - \log_{10}(U(t)/Ir(t))]$$
$$= 14.8\text{Gyr}[-1.30 + 2.10]$$
$$= 11.8\text{Gyr}.$$

Taking the error bars into account, the authors deduced an age for ^{238}U of 12.5 billion years, give or take 3 billion years. This is the age of the heavy element and is therefore the time since what was probably the earliest supernova explosion. When increased by 0.1–0.3 billion years to account for the time between the creation of the universe and the earliest supernova to form as the end stage of a star's life cycle,

Figure 4.2. Professor Willard F. Libby (1908–1980). (©The Nobel Foundation.)

these estimates give the minimum age of the universe to be around 13 billion years old.

4.5 Carbon Dating

Another, more common, method for dating objects on earth is called carbon dating, which was developed in the late 1940s by an American chemist, W. F. Libby, at the University of California at Los Angeles (Figure 4.2). For this work he received the Nobel Prize in Chemistry in 1960.

Let us introduce the application of this method to the problem of authenticating King Arthur's Round Table, as discussed in the book *Applying Mathematics*, by D. N. Burghes, I. Huntley, and J. McDonald (1982).

In the great hall in Winchester Castle there is a big round table made of oak that is 18 feet in diameter and divided into 25 sections, one for the king and the other sections for the 24 knights whose names are inscribed (Figure 4.3). It fits the legend of King Arthur's Round Table from the 5th century. The table was known to be at least several hundred years old, as the medieval historian John Harding reported in his *Chronicle* (1484) that the round table "began at Winchester and ended at Winchester, and there it hangs still."

To put an end to the speculation about the Winchester round table, in 1976 it was taken down from the wall on which it was hung, and

Figure 4.3. The Winchester round table. Winchester was the presumed site of Camelot. (Used by permission of Hampshire County Council.)

a series of tests were done to determine the age of the table. Carbon dating was one of the methods used.

Earth's atmosphere is continuously bombarded by cosmic rays. This produces neutrons in the atmosphere, and these neutrons combine with nitrogen, forming carbon-14 (^{14}C). (See Figure 4.4.) Living plants "breathe in" CO_2, along with $^{14}CO_2$; the rate of absorption of ^{14}C is balanced by the natural decay, and an equilibrium is reached. When the sample is formed (in this case, when the table was made) the wood is isolated from its original environment and the ^{14}C atoms decay without any further absorption. This behavior of (dead) wood, i.e., wood whose ^{14}C atoms decay without being replenished, is governed by Eq. (4.1) or (4.2). The rate of decay at present (1977) was measured with the sample taken from the table. It was found that $R(t) \equiv -dN/dt = 6.08$ per minute per gram of sample.

Using Eqs. (4.1) and (4.2), we know that the rate of decay $R(t)$ is, theoretically,

$$R(t) \equiv -dN/dt = \lambda e^{-\lambda t} N_0 = e^{-\lambda t} R(0),$$

where $R(0) = \lambda N_0$ would be known if we knew the original concentration of ^{14}C in the wood, which we don't!

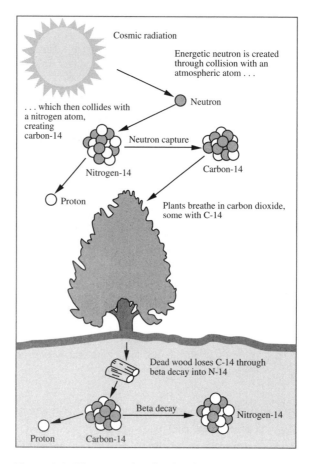

Figure 4.4. The natural cycle of carbon-14.

Here is the crucial *assumption* in carbon dating: we can measure $R(0)$ at present from living plants. That is, the rate of cosmic production of ^{14}C and its decay in living plants in the 5th century is assumed to be approximately the same as the rate in living plants today. By measuring living wood, we find, by this argument, $R(0) = 6.68$ per minute per gram of sample. Thus

$$t = \frac{1}{1.245 \times 10^{-4}} \ln \frac{6.68}{6.08} = 756 \text{ years.}$$

This gives a date for the table of about AD 1220, which clearly indicates that the table was not King Arthur's (he lived in the 5th century).

4.6 HIV Modeling

During the early days of AIDS (acquired immunodeficiency syndrome) research, it was not understood that, in individuals infected with HIV (human immunodeficiency virus), the virus was being produced at a prolific rate even before the onset of symptoms and full-blown AIDS. The modeling effort of Perelson et al. (1996) quantified this viral production rate and did much to advance the treatment strategies for AIDS by David Ho and others.

Perelson et al.'s model was incredibly simple. The rate of change of $V(t)$, the concentration of viral particles (called virions) in blood plasma, is given by

$$\frac{dV}{dt} = P - cV, \tag{4.5}$$

where P is the rate of production of new HIV virions and c is the "clearance" rate for the virions in the plasma. The virus is eliminated continually by the body's own immune cells or by natural body clearance functions, even in the absence of any drug therapy. Both $P(t)$ and c are, however, unknown.

A protease inhibitor, *ritonavir*, was administered orally to five infected patients. After treatment, the HIV virus concentrations in plasma were measured at very high frequency (every two hours until the sixth hour, every six hours until day 2, and every day until day 7).

It was found that each patient responded with a similar pattern of viral decay (see Figure 4.5). Assuming that the drug killed the production of new virus completely, P was set to zero. Equation (4.5) becomes

$$\frac{dV}{dt} = -cV, \tag{4.6}$$

which was solved to yield

$$V(t) = V(t_0)e^{-c(t-t_0)} \quad \text{for } t > t_0, \tag{4.7}$$

where t_0 is the time when the effect of the drug took hold, a day or so after treatment started.

By plotting $\ln(V(t)/V(t_0))$ against $t - t_0$ and using linear regression to determine the slope, the authors found a mean half-life of the virus in the plasma of

$$\tau = \ln2/c \cong 0.18 \text{ to } 0.34 \text{ days.}$$

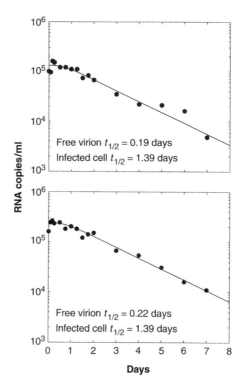

Figure 4.5. Log of plasma concentrations (copies per mL) of HIV-1 RNA (circles) for two representative patients (upper panel, patient 104; lower panel, patient 107) after *ritonavir* treatment was begun on day 0. The solid line is a nonlinear least square fit to the data. HIV-1 RNA level is an easier measure of HIV virions since each HIV virion contains two RNA molecules. (See exercise 5 for more details.) (From Perelson et al. [1996], used by permission of Alan S. Perelson.)

Thus c is determined to be 2.06 to 3.81 per day. $V(t_0)$ was also measured as the viral concentration before the drugs took effect:

$$V(t_0) \sim 3 \times 10^5 \text{ virions per mL of plasma.}$$

To find P, the authors assumed that before the administration of drugs, there was a quasi-steady state when viral production was balanced by viral clearance. (In fact, only patients not yet experiencing the onset of full-blown AIDS, when viral production overwhelms the body's ability to clear it, were chosen for the study.) For this quasi-steady state, $\frac{d}{dt} V$ was set to zero in Eq. (4.5), leading to

$$P - cV = 0, \text{ for } t \leq t_0. \tag{4.8}$$

From this, they found

$$P(t_0) \cong c V(t_0) \sim 2 \times 3 \times 10^5$$

virions per mL of plasma per day.

That amounts to almost a billion new viral particles a day produced in each liter of blood during what, at the time, was thought of as the "dormant" phase of AIDS! It turns out that even in the early stages of HIV infection, the virus was being produced at an incredible rate ("the raging fire of active HIV replication"). The body was able to clear the virus out at a rapid rate also, until it could not keep up any longer. Based on this work, the authors suggested that "early and aggressive therapeutic intervention is necessary if a marked clinical impact is to be achieved."

A review article for applied mathematicians can be found in Perelson and Nelson (1999).

4.7 Exercises

1. Christ and the Disciples at Emmaus

In the 1930s, the painting *Christ and the Disciples at Emmaus* (Figure 4.6) was certified as a genuine 17th century Vermeer by a noted art historian, A. Bredius, and bought by the Rembrandt Society. In 1945, an art forger, Han van Meegeren, announced in a Belgian prison that he was the painter of *Disciples at Emmaus*, an admission made presumably to avoid prosecution on the charge that he had sold actual Vermeer paintings to Nazis during the war.

A pigment of major importance in paintings is white lead, which contains a radioactive isotope, ^{210}Pb. It is manufactured from ores that contain uranium and elements to which uranium decays. One of these elements is radium-226 (^{226}Ra), which has a half-life of 1,600 years, and decays to ^{210}Pb, which has a half-life of 22 years. While still part of the ore, the amount of ^{226}Ra decaying to ^{210}Pb is equal to the amount of ^{210}Pb disintegrating into some other element. That is, ^{226}Ra and ^{210}Pb are in a "radioactive equilibrium."

In the manufacture of the pigment, the radium and most of its descendants are removed. The ^{210}Pb begins to decay without replenishment.

Let $y(t)$ be the number of ^{210}Pb atoms per gram of ordinary lead at time t. Let t_0 be the time the pigment was manufactured and r the number of disintegrations of ^{226}Ra per gram of ordinary lead per unit time.

Figure 4.6. *Christ and the Disciples at Emmaus.*

a. Explain why the following equations should govern the change in the amount of ^{210}Pb:

$$\frac{dy}{dt} = -\lambda y + r \quad \text{while in the ore,} \tag{4.9}$$

$$\frac{dy}{dt} = -\lambda y \quad \text{after manufacture.} \tag{4.10}$$

λ is the decay constant for ^{210}Pb.

b. Measurements from a variety of ores over the earth's surface gave a range of values for the rate of disintegration of ^{226}Ra per gram of ordinary lead as

$$r = 0 - 200 \text{ per minute.}$$

Show that it is reasonable to assume that

$$\lambda y(t_0) = r = 0 - 200 \text{ per minute.}$$

c. Solve (4.10) subject to the initial condition

$$y(t_0) = r/\lambda.$$

d. For the *Disciples at Emmaus* painting, it was measured that

$$-\frac{dy}{dt}(t) \cong 8.5 \text{ per minute.}$$

Estimate $t - t_0$ to decide if the painting can be 300 years old.

2. *Lascaux Cave paintings*

Charcoal from the dwelling level of the Lascaux Cave in France gives an average count of 0.97 disintegrations of ^{14}C per minute per gram of sample. Living wood gives 6.68 disintegrations per minute per gram. Estimate the date of occupation and hence the probable date of the wall painting in the Lascaux Cave.

3. *Age of uranium*

The currently measured value of $\log_{10}(U/Th)$ in star CS31082-001 is -0.74 ± 0.15. U denotes the concentration of the ^{238}U isotope, and Th that of ^{232}Th, which has a half-life of 14 Gyr. Because of the proximity of the two elements in their atomic mass numbers, their initial ratio, $(U/Th)_0$, at the time of their nuclearsynthesis is less affected by theoretical uncertainties. The theory gives $\log_{10}(U/Th)_0 = -0.10$. Derive the age of uranium in that ancient star, and hence the minimum age of the universe.

4. *A slightly more involved HIV model*

Cells that are susceptible to HIV infection are called T (target) cells. Let $T(t)$ be the population of uninfected T-cells, $T^*(t)$ that of the infected T-cells, and $V(t)$ the population of the HIV virus. A model for the rate of change of the infected T-cells is

$$\frac{dT^*}{dt} = kVT - \delta T^*, \tag{4.11}$$

where δ is the rate of clearance of infected cells by the body, and k is the rate constant for the infection of the T-cells by the virus. The equation for the virus is the same as Eq. (4.5):

$$\frac{dV}{dt} = P - cV, \tag{4.12}$$

but now the production of the virus can be modeled by

$$P(t) = N\delta T^*(t).$$

Here N is the total number of virions produced by an infected T-cell during its lifetime. Since $1/\delta$ is the length of its lifetime, $N\delta T^*(t)$ is the total rate of production of $V(t)$.

At least during the initial stages of infection, T can be treated as an approximate constant. Equations (4.11) and (4.12) are the two coupled equations for the two variables $T^*(t)$ and $V(t)$.

A drug therapy using RT (reverse transcriptase) inhibitors blocks infection, leading to $k \cong 0$. Setting $k = 0$ in (4.11), solve for $T^*(t)$. Substitute it into (4.12) and solve for $V(t)$. Show that the solution is

$$V(t) = \frac{V(0)}{c - \delta}[ce^{-\delta t} - \delta e^{-ct}].$$

5. Protease inhibitors

A drug therapy using protease inhibitors causes infected cells to produce noninfectious virions. It becomes necessary in our model to separate the infectious virion population $V_I(t)$ from the noninfectious virion population $V_{NI}(t)$, with $V(t) = V_I(t) + V_{NI}(t)$. Equations (4.11) and (4.12) remain valid except with V replaced by $V_{NI}(t)$:

$$\frac{d}{dt}T^* = kV_I T - \delta T^*, \tag{4.13}$$

$$\frac{d}{dt}V_{NI} = N\delta T^* - cV_{NI}. \tag{4.14}$$

The equation for $V_I(t)$ is given by

$$\frac{d}{dt}V_I = -cV_I \tag{4.15}$$

because infectious virions are no longer produced with an effective protease inhibitor treatment but are only being cleared by the body at the rate c.

a. Solve Eq. (4.15), substituting it into Eq. (4.13) to show that the solution for $T^*(t)$ is, assuming $T = T_0$ is a constant,

$$T^*(t) = T^*(0)e^{-\delta t} + \frac{kT_0 V_0(e^{-ct} - e^{-\delta t})}{\delta - c}$$

$$= kV_0 T_0[ce^{-\delta t} - \delta e^{-ct}]/[\delta(c - \delta)].$$

The last step is obtained by assuming that T^* is in a quasi-steady state before the therapy (set $\frac{d}{dt}T^* = 0$ at $t = 0$ in (4.13), with $T = T_0$ and $V_I = V_0$).

b. Substitute $T^*(t)$ found in (a) into (4.14) to show:

$$V_{NI}(t) = \frac{cV_0}{c - \delta}\left[\frac{c}{c - \delta}(e^{-\delta t} - e^{-ct}) - \delta t e^{-ct}\right].$$

c. Adding $V_{NI}(t)$ and $V_I(t)$, show that the total virion concentration is given by

$$V(t) = V_0 e^{-ct} + \frac{cV_0}{c - \delta}\left[\frac{c}{c - \delta}(e^{-\delta t} - e^{-ct}) - \delta t e^{-ct}\right].$$

Since measurements cannot distinguish between infectious and noninfectious virions, it is the total virions that are measured in Figure 4.5. There are two decay rates, c and δ. These are obtained by best fitting the observed decay of $V(t)$ with the above solution, using δ and c as the two parameters.

5

Modeling in the Physical Sciences, Kepler, Newton, and Calculus

Mathematics required:
        calculus, vectors, Cartesian and polar coordinates

5.1 Introduction

Nowadays classical mechanics is a deductive science: we know the equations of motion and we use them to *deduce* results. We no longer "model" the orbits of the planets, for example. We *model* biological, psychological, ecological, and social problems because in these fields the governing equations are not yet known.

Astronomy, in particular the prediction of planetary orbits, was an empirical exercise in Johannes Kepler's time (1571–1630), just as biology is to us in modern times. Kepler did not know the law of gravitation. Calculus had not even been invented. Both had to await Isaac Newton (1642–1727). Newton's law of universal gravitation has been regarded as one of the greatest achievements of human thought. But, as Newton freely admitted, he was standing on the shoulders of those before him. If we take *you* back in time, to Kepler's time, but equip you with the mathematics of calculus, can you *model* the empirical data Kepler had in his possession at the time and beat Newton to arrive at the theory of gravitation? This is modeling at its best: to come up with an equation to explain the observational data, to generalize to other situations, to make new predictions, and to verify the new theory with additional experiments.

Johannes Kepler (Figure 5.1) was born near Stuttgart, Germany, and raised mostly by his mother. An infection from smallpox at age 4 left his eyesight much impaired. He was educated in theology and mathematics at the Lutheran seminary at the University of Türbingen, and in 1594 accepted a lectureship at the University of Graz in Austria. In 1600 he became the assistant to the famous Danish-Swedish astronomer Tycho Brahe, the court astronomer to Kaiser Rudolph II at Prague. Prior to this time, Kepler's work had been more metaphysical—his first treatise was entitled *The Cosmic Mystery*. Among other things Greek, Kepler

Figure 5.1. Johannes Kepler (1571–1630).

thought the Golden Ratio was a fundamental tool of God in creating the universe.

Johannes Kepler was a mathematician rather than an observer; he viewed his study of astronomy as fulfilling his Christian duty to understand God's creation, the universe. Kepler, however, had the great fortune to have inherited, upon Brahe's sudden death in 1601, both his master's post as the imperial mathematician and his large collection of accurate data on the motion of Mars. Based on these data and after toiling for two decades, Kepler formed the following three laws on the orbits of planets:

I. The planets move about the sun in elliptical orbits with the sun at one focus.

II. The areas swept over by the radius vector drawn from the sun to a planet in equal times are equal.

III. The squares of the times of describing the orbits (periods) are proportional to the cubes of the major axes.

These three laws (Figures 5.2 and 5.3) represent a distillation of volumes of empirical data compiled by Brahe, which was the most time-consuming part of the whole modeling process. Kepler had been driven and sustained by the conviction that: "By the study of the orbit of Mars, we must either arrive at the secrets of astronomy or forever remain in ignorance of them." Kepler had first tried the circle, believed by the Greeks to be the perfect orbit that heavenly bodies must follow, before settling on the ellipse. It is a tribute to the accuracy of Brahe's

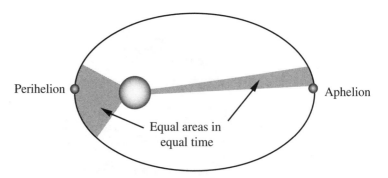

Figure 5.2. Kepler's first and second laws. The sun sits in one of the foci of an ellipse, which forms a planet's orbit. The planet moves fastest when it is closest to the sun (in the orbit's perihelion) and slowest when it is farthest away (in the orbit's aphelion).

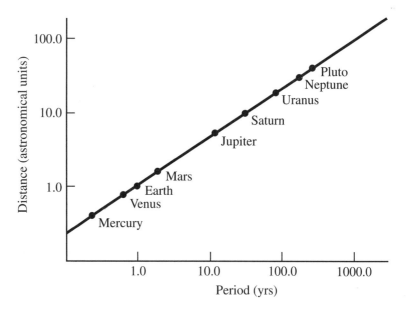

Figure 5.3. A schematic log-log plot of the semimajor axis of a planet and its period around the sun, from Kepler's third law. (Courtesy of Vik Dillion, University of Sheffield.)

data that Kepler could distinguish a circle or an oval from an ellipse in each's fit to the observed orbit of Mars. The first and the second laws (see Figure 5.2) were published in 1609. The third law (Figure 5.3) was not found until 10 years later. Some thought that the invention of logarithms by John Napier (1614) played some role in helping Kepler visualize the 1.5 exponent present in his third law.

Figure 5.4. Sir Isaac Newton (1642–1727).

It turns out that Kepler's three laws already contained the information needed to arrive at the law of gravitation; it needed an additional step of distillation and mathematical modeling, and the latter required a knowledge of calculus. Since you know calculus, let us see if you can model Kepler's three laws by a differential equation and possibly arrive at Newton's law of gravitation. The presentation here is the reverse of what you may find in other texts, where Newton's law of gravitation is assumed axiomatically and Kepler's laws are then *derived*.

5.2 Calculus, Newton, and Leibniz

Calculus is probably your first introduction to college mathematics. Instead of dealing with finite numbers, you encounter "infinitesimals," such as dx and dy, nonintuitive quantities that are infinitely small and yet nonzero. Using it, you are able to find the slope of a curve $y = f(x)$ at any point x as $\frac{dy}{dx}$, and the instantaneous velocity $u(t)$ of a particle whose displacement is $x(t)$ as the derivative $u(t) = \frac{d}{dt}x(t)$. This concept then becomes indispensable in physics, where you need to calculate instantaneous velocity $u(t)$ and acceleration $a(t) = \frac{d}{dt}u(t)$. Using integration calculus you are also able to calculate the area or volume of curved objects.

The foundation of calculus was laid in the 17th century by Isaac Newton, the English scientist better known for his work in physics (Figure 5.4), and the German mathematician Gottfried Leibniz (1646–1716). It appears that the two developed calculus independently. Newton probably started work on what he called the "method of fluxions" earlier, in 1665–1666, and even wrote up the paper in 1671, but failed to get it published until 1736. Leibniz published details of his differential calculus in 1684 and integral calculus in 1686. We owe our modern notation of dy/dx and $\int dx$ to Leibniz in these two publications. Although Leibniz published his results first, he had to defend himself in the last years of his life against charges of plagiarism by various members of the Royal Society of London. It was one of the longest and bitterest priority disputes in the history of mathematics and science and was fought mainly along nationalistic lines. In response to Leibniz's protests, the Royal Society of London set up a committee to investigate and decide on the priority. The committee report found in favor of Newton, who, as the president of the Royal Society, actually wrote the report.

Newton, as a physicist, was interested in finding the instantaneous velocity of a particle tracing a path whose coordinate is $(x(t), y(t))$ as time t changes. He denoted the velocity as $(\dot{x}(t), \dot{y}(t))$, which Newton called the "fluxions" of $(x(t), y(t))$ associated with the flux of time. The "flowing quantities" themselves, $(x(t), y(t))$, are called "fluents." To find the tangent to the curve traced by $(x(t), y(t))$, Newton used the ratio of the finite quantities $\dot{y}$ and $\dot{x}$, as $\dot{y}/\dot{x}$. Leibniz, on the other hand, introduced dx and dy as infinitesimal differences in x and y, and obtained the tangent dy/dx as the finite ratio of two infinitesimals. For integration, Newton's method involves finding the fluent given a fluxion, while Leibniz's treats integration as a sum. The way calculus is taught nowadays more closely resembles the concepts of Leibniz than Newton, although modern textbooks still vacillate between treating integration as antidifferentiation (a Newtonian concept) and as a Riemann sum (similar to Leibniz's idea).

5.3 Vector Calculus Needed

Let

$$\mathbf{r} = x\hat{\mathbf{i}} + y\hat{\mathbf{j}}$$

be the position vector, where $\hat{\mathbf{i}}$ and $\hat{\mathbf{j}}$ are the unit vectors in the x and y directions, respectively. They are constant (Cartesian) vectors.

Let θ be the angle $\mathbf{r}$ makes relative to the x-axis and r be the magnitude of $\mathbf{r}$. Then (see Figure 5.5):

$$x = r\cos\theta, \quad y = r\sin\theta.$$

Let $\hat{\mathbf{r}}$ and $\hat{\boldsymbol{\theta}}$ be the unit vectors in the r and θ directions, respectively. They can be expressed in terms of $\hat{\mathbf{i}}$ and $\hat{\mathbf{j}}$ as

$$\hat{\mathbf{r}} = \cos\theta\hat{\mathbf{i}} + \sin\theta\hat{\mathbf{j}}, \quad \hat{\boldsymbol{\theta}} = -\sin\theta\hat{\mathbf{i}} + \cos\theta\hat{\mathbf{j}}.$$

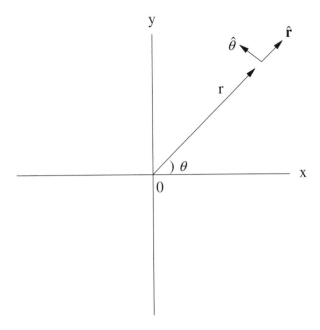

Figure 5.5. Cartesian and polar coordinates.

These (non-Cartesian) vectors may change direction in time although their magnitudes are defined to be always 1. In terms of the polar coordinates, the position vector can be written as

$$\mathbf{r} = r\cos\theta\hat{\mathbf{i}} + r\sin\theta\hat{\mathbf{j}} = r(\cos\theta\hat{\mathbf{i}} + \sin\theta\hat{\mathbf{j}})$$

$$= r\hat{\mathbf{r}}.$$

If $\mathbf{r}$ is the position of a particle, then its velocity, $\mathbf{v}$, is given by $d\mathbf{r}/dt$, the rate of change of $\mathbf{r}$:

$$\mathbf{v} = \frac{d}{dt}\mathbf{r} = \frac{d}{dt}(r\hat{\mathbf{r}}) = \frac{dr}{dt}\hat{\mathbf{r}} + r\frac{d}{dt}\hat{\mathbf{r}}.$$

Since

$$\frac{d}{dt}\hat{\mathbf{r}} = \frac{d}{dt}(\cos\theta\hat{\mathbf{i}} + \sin\theta\hat{\mathbf{j}}) = -\sin\theta\frac{d\theta}{dt}\hat{\mathbf{i}} + \cos\theta\frac{d\theta}{dt}\hat{\mathbf{j}}$$

$$= \frac{d\theta}{dt}(-\sin\theta\hat{\mathbf{i}} + \cos\theta\hat{\mathbf{j}}) = \frac{d\theta}{dt}\hat{\boldsymbol{\theta}}$$

(and, for later use, $\frac{d}{dt}\hat{\boldsymbol{\theta}} = \frac{d}{dt}(-\sin\theta\hat{\mathbf{i}} + \cos\theta\hat{\mathbf{j}}) = (-\cos\theta\hat{\mathbf{j}} - \sin\theta\hat{\mathbf{j}})\frac{d\theta}{dt} = -\frac{d\theta}{dt}\hat{\mathbf{r}}$), we have

$$\mathbf{v} = \dot{r}\hat{\mathbf{r}} + r\dot{\theta}\hat{\boldsymbol{\theta}},$$

where we have used an overhead dot to denote d/dt, following Newton. The d/dt notation was due to Leibniz.

Since acceleration is the rate of change of velocity, we have

$$\mathbf{a} = \frac{d}{dt}\mathbf{v} = \frac{d}{dt}(\dot{r}\hat{\mathbf{r}} + r\dot{\theta}\hat{\boldsymbol{\theta}})$$

$$= \ddot{r}\hat{\mathbf{r}} + \dot{r}\dot{\theta}\hat{\boldsymbol{\theta}} + r\ddot{\theta}\hat{\boldsymbol{\theta}} + \dot{r}\frac{d}{dt}\hat{\mathbf{r}} + r\dot{\theta}\frac{d}{dt}\hat{\boldsymbol{\theta}}$$

$$= (\ddot{r} - r\dot{\theta}^2)\hat{\mathbf{r}} + (r\ddot{\theta} + 2\dot{r}\dot{\theta})\hat{\boldsymbol{\theta}}$$

$$\equiv a_r\hat{\mathbf{r}} + a_\theta\hat{\boldsymbol{\theta}},$$

where we have defined

$$a_r \equiv \ddot{r} - r\dot{\theta}^2 \tag{5.1}$$

as the component of acceleration in the $\hat{\mathbf{r}}$ direction, and

$$a_\theta \equiv r\ddot{\theta} + 2\dot{r}\dot{\theta} \tag{5.2}$$

as the component of acceleration in the $\hat{\boldsymbol{\theta}}$ direction.

5.4 Rewriting Kepler's Laws Mathematically

Kepler's three laws are mostly about geometry. Only the second law hints at some rates of change of angle.

If over time Δt, the orbit of a planet describes an arc with an extended angle $\Delta\theta$, the area swept over by the radius vector (measured from the sun to the planet) in Δt is, from geometry,

$$\Delta A = \frac{1}{2}r^2\Delta\theta.$$

This result is obtainable by assuming that for very small $\Delta\theta$, there is no difference between an elliptical sector and a circular sector. For the latter, the area of a sector is $\frac{\Delta\theta}{2\pi}$ of πr^2, which is $\frac{1}{2}r^2\Delta\theta$. Now you are beginning to appreciate the advantages of being able to examine infinitesimal changes afforded by calculus.

Kepler's second law says:

$$\Delta A/\Delta t \text{ is a constant.}$$

Thus

$$\frac{1}{2}r^2\Delta\theta/\Delta t \text{ is a constant.}$$

Taking the limit $\Delta t \to 0$, we get

$$r^2 \dot{\theta} = h, \quad \text{a constant,} \qquad (5.3)$$

since $\dot{\theta} = \frac{d}{dt}\theta = \lim \Delta\theta / \Delta t$. Equation (5.3) is a mathematical statement of Kepler's second law. (This is actually a statement of the conservation of angular momentum in Newtonian mechanics, but Kepler was not aware of its general nature.) It can also be stated as

$$\frac{d}{dt}(r^2\dot{\theta}) = \frac{d}{dt}(h) = 0. \qquad (5.4)$$

Since the acceleration in the $\hat{\theta}$ direction is

$$a_\theta = r\ddot{\theta} + 2\dot{r}\dot{\theta}, \quad \text{then } r a_\theta = \frac{d}{dt}(r^2\dot{\theta}).$$

Equation (5.4) implies that

$$a_\theta = 0, \quad r > 0. \qquad (5.5)$$

Stated this way, Kepler's second law says that the acceleration of the planet is purely in the radial direction, likely towards the sun. This is the first hint of a force of attraction between the planet and the sun. Next we shall rewrite the radial acceleration in a mathematical form.

Kepler's first law states that the orbit of a planet can be described by an ellipse. In terms of its radial distance from the sun, the formula for an ellipse is

$$r = \frac{p}{1 + e\cos\theta}, \quad p > 0, \quad 0 < e < 1. \qquad (5.6)$$

This is a standard formula for an ellipse whose semimajor axis is given by $p/(1 - e^2)$.

From Eq. (5.1), the acceleration in the radial direction is

$$a_r = \ddot{r} - r\dot{\theta}^2.$$

We next proceed to find a_r by differentiating Eq. (5.6) with respect to time:

$$\dot{r} = \frac{ep\sin\theta\,\dot{\theta}}{(1 + e\cos\theta)^2} = \frac{e}{p}r^2\dot{\theta}\sin\theta = \frac{eh}{p}\sin\theta,$$

using Eq. (5.3) in the last step. Continuing,

$$\ddot{r} = \frac{eh}{p}\cos\theta\dot{\theta} = \frac{eh^2}{pr^2}\cos\theta;$$

again Eq. (5.3) was used in the last step. Also,

$$r\dot{\theta}^2 = r^4\dot{\theta}^2/r^3 = h^2/r^3.$$

Thus,

$$a_r = \ddot{r} - r\dot{\theta}^2 = -\frac{h^2}{r^2}\left[\frac{1}{r} - \frac{e\cos\theta}{p}\right]$$

$$= -\frac{h^2}{r^2}\left[\frac{1+e\cos\theta}{p} - \frac{e\cos\theta}{p}\right].$$

Finally, with the cancellation of the cosine terms:

$$a_r = -\left(\frac{h^2}{p}\right)\frac{1}{r^2}. \tag{5.7}$$

This is the inverse square law! The acceleration is inversely proportional to the square of the radial distance and it is negative, i.e., towards the sun.

Equation (5.7) is not yet a "universal law" because h and p are characteristics of individual planets. Kepler's third law remedies this situation. We are fortunate that Kepler did not stop at his first two laws but worked an extra 10 years to get his third law in order. (It turns out that Kepler's third law is only approximately correct, and needs a correction if the sun and the planet are of comparable mass. This does not matter much for our purpose, though.)

Let R be the semimajor axis of an ellipse. Then $R = p/(1 - e^2)$. Let T be the period, the time it takes for the planet to make a complete revolution around the sun. Kepler's third law states that

$$T^2/R^3 = \text{the same constant} \tag{5.8}$$

for any planet around the same sun.

From Eq. (5.3) we have

$$r^2\frac{d\theta}{dt} = h \quad \text{or} \quad dt = r^2 d\theta/h.$$

We can integrate both sides to yield the period $T \equiv \int_0^T dt$:

$$T = \frac{1}{h} \int_0^{2\pi} r^2 d\theta = \frac{p^2}{h} \int_0^{2\pi} \frac{d\theta}{(1 + e\cos\theta)^2} = \frac{p^2}{h} \cdot \frac{4\pi}{(1 - e^2)^{3/2}},$$

using tables of integrals to do the last trigonometric integral. A better way is to use complex variables, but some readers may not have learned that yet.

Now, since

$$T^2 = \frac{(4\pi)^2 p^4}{h^2(1 - e^2)^3} \quad \text{and} \quad R^3 = \frac{p^3}{(1 - e^2)^3},$$

$$T^2/R^3 = (4\pi)^2 p/h^2.$$

Kepler's third law, Eq. (5.8), then says that $K \equiv h^2/p$ must be the same constant regardless of which planet we are considering.

The inverse square law, Eq. (5.7), becomes

$$\boxed{a_r = -\frac{K}{r^2},} \tag{5.9}$$

where K is *independent of the planet's properties*. This is almost a "universal" law!

5.5 Generalizations

Using calculus, we distilled from Kepler's three laws the following results: the acceleration of the planet is purely in the radial direction towards the sun and is given by Eq. (5.9). The constant K is independent of the properties of the planet under consideration, but may depend on the properties of the sun. After all, all the observations of Brahe and Kepler were for planets in our solar system. To find out what K is, we need to go beyond Kepler. The process of reasoning is not difficult at all.

It is reasonable to assume that

    a. the sun is the source of this "force of attraction"—the cause for the radial acceleration of the planet towards it; and

    b. this attraction should be stronger for a sun that is more massive.

1. Let us start with the "working hypothesis" that K is proportional to the mass of the sun M, i.e.,

$$K = GM,$$

where G is the proportionality constant. G should not depend on the properties of the sun, nor should it depend on the properties of the planet. So G should be a "universal" constant. Thus

$$a_r = -\frac{GM}{r^2}. \tag{5.10}$$

Equation (5.10) is, essentially, the Universal Law of Gravitation, and we have just derived it! (The form you probably have seen is $F_r = -\frac{GmM}{r^2}$, for the force on a planet of mass m. It is equivalent to Eq. (5.10), since $F_r = ma_r$.)

2. How do we determine G? Let's wait until later. Newton didn't know the value of G either and never figured it out, except that he knew it must be very small.

3. Newton later called this "force of attraction" *gravity* and postulated that it exists not just between a planet and the sun but between the moon and the earth as well. (The nature of such a force was not clarified until Einstein, three centuries later.) Galileo (1564–1642) had observed the four brightest moons of Jupiter in orbit around Jupiter just as the planets revolve around the sun. It was known in Kepler's time that our moon revolves around the earth in an elliptical orbit with a very small eccentricity. Applying Eq. (5.10) to the moon, we get

$$a_r = -\frac{GM_{\text{Earth}}}{r^2}, \quad \text{towards the earth.}$$

4. This gravitational pull by the earth should apply to "falling apples" and "cannonballs" as well as the moon. In the case of falling objects near the surface of the earth, it was determined that they accelerate downward (towards the center of the earth) with magnitude

$$|a_r| = g = 980\,\text{cm/s}^2.$$

So, letting $r = a + z$, where a is the radius of the earth ($a \cong 6{,}400$ km) and z is the height above the surface of the earth, we have

$$g = \frac{GM_{\text{Earth}}}{r^2} = \frac{GM_{\text{Earth}}}{(a+z)^2} \cong \frac{GM_{\text{Earth}}}{a^2}, \quad \text{for } z/a \ll 1.$$

This way, G is determined from

$$GM_{Earth} = a^2 g = 4 \times 10^{20}\,\text{cm}^3\,\text{s}^{-2}.$$

5. If we can somehow "weigh" the earth and find that $M_{Earth} \cong 6 \times 10^{27}$ gm, we will eventually get

$$G = 0.67 \times 10^{-7}\,\text{cm}^3 \cdot \text{s}^{-2} \cdot \text{gm}^{-1}.$$

The experiment to "weigh the earth" was not carried out until Henry Cavendish (1731–1810) did it in 1797–1798, several decades after Newton's death. However, even without knowing G explicitly, many interesting results can be obtained, as in the first three of the exercises at the end of this chapter.

5.6 Newton and the Elliptical Orbit

In August 1684, the English astronomer Edmond Halley (1656–1742) visited Isaac Newton in Cambridge and asked him if he knew the shape of the orbit of a planet subjected to a force that is proportional to the reciprocal of the square of the distance from the sun. Newton was able to say in reply that he had solved this problem and it was an ellipse, though the result had not yet been published. At Halley's urging and expense, Newton published his celebrated *Principia Mathematica* in 1687, where the three laws of mechanical motion were stated as axioms along with the law of universal gravitation. If you take the axioms as given, the elliptical orbits will follow inevitably. The solution is outlined below. Note that the *solution* procedure is the reverse of the *modeling* procedure.

From the second law of motion, the acceleration of an object is equal to the force (per unit mass) acting on it. So in the case of gravitational force, we have, for the planet

$$a_r = -\frac{GM}{r^2}, \tag{5.11}$$

$$a_\theta = 0,$$

where M is the mass of the sun and r is the distance of the planet from it. From Eq. (5.1), we know that $a_r = \ddot{r} - r\dot{\theta}^2$ and $a_\theta = r\ddot{\theta} + 2\dot{r}\dot{\theta}$. Thus

$$\ddot{r} - r\dot{\theta}^2 = -\frac{GM}{r^2}, \tag{5.12}$$

$$r\ddot{\theta} + 2\dot{r}\dot{\theta} = 0. \tag{5.13}$$

From Eq. (5.13) we know that $r^2\dot\theta = h$ is constant, which is specified by the initial velocity. Substituting this result into Eq. (5.12), we obtain an equation involving $r(t)$ exclusively:

$$\ddot r - \frac{h^2}{r^3} = -\frac{GM}{r^2}. \tag{5.14}$$

To obtain the shape of the orbit (i.e., r as a function of θ), we make use of $d\theta = \frac{h}{r^2}dt$, so that

$$\ddot r = \frac{d^2}{dt^2}r = \frac{d}{dt}\frac{d\theta}{dt}\frac{dr}{d\theta} = \frac{h}{r^2}\frac{d}{d\theta}\left(\frac{h}{r^2}\frac{d}{d\theta}r\right) = -\frac{h^2}{r^2}\frac{d^2}{d\theta^2}\left(\frac{1}{r}\right).$$

Consequently, Eq. (5.14) becomes

$$\frac{d^2}{d\theta^2}\left(\frac{1}{r}\right) + \frac{1}{r} = \frac{GM}{h^2}. \tag{5.15}$$

This is a linear second-order ordinary differential equation for $(\frac{1}{r})$ with constant forcing. The solution is

$$\frac{1}{r} = A\cos(\theta - \theta_0) + \frac{GM}{h^2}. \tag{5.16}$$

The two constants, A and θ_0, can be determined by the initial conditions on the position. The solution in Eq. (5.16) is in the form of an ellipse, (5.6).

5.7 Exercises

1. How far is the moon?

The moon's orbit around the earth is nearly circular, with a period of 28 days. Determine how far the moon is from (the center) of the earth using only this and the following information, which was available to Newton.

a. The gravitational acceleration of the moon caused by the gravitational pull of the earth is

$$GM_{Earth}/r^2,$$

where M_{Earth} is the mass of the earth (but the values of G or M_{Earth} were not known separately).

b. The orbit of the moon (or any satellite, for that matter) is a balance between the gravitational acceleration and the centrifugal force v^2/r. That is,

$$GM_{\text{Earth}}/r^2 = v^2/r,$$

where $v = r\frac{d\theta}{dt}$ is the velocity in the angular direction.

c. The gravitational acceleration for an object near the earth's surface ($r \cong a = 6{,}400$ km) is known to be $g = 980$ cm/s^2.

2. The mass of the sun

Determine the mass of the sun in units of the earth mass (i.e., find $\overline{M} = M_{\text{sun}}/M_{\text{Earth}}$) using only the information provided in exercise 1 and the following information:

a. the period of earth's orbit is 1 year.

b. the sun–earth distance is, on average, about 1.5×10^8 km.

3. Geosynchronous satellite

If you want to put a satellite in a geosynchronous orbit (so that the satellite will always appear to be above the same spot on earth), how high (measured from the center of the earth) must it be placed? You are given $GM_E = a^2 g = 4 \times 10^{20}$ cm^3 s^{-2}.

4. Weighing a planet

Newton's law of gravitation and the requirement that the centrifugal acceleration of a body revolving around a planet should be equal to the gravitational pull of the planet suggest a way to determine the mass M of any planet with a satellite or moon. For this exercise you need to know $G = 0.67 \times 10^{-7}$ cm^3 s^{-2} gm^{-1}.

Determine the mass of a planet when you know only that its moon revolves around it with a nearly circular orbit with $r = 4 \times 10^5$ km once every 30 days.

5. Weighing Jupiter

Find the mass of Jupiter, given that its moon Callisto has a mean orbital radius of 1.88×10^6 km and an orbital period of 16 days and 16.54 hours. For this exercise you need to know G.

6

Nonlinear Population Models: An Introduction to Qualitative Analysis Using Phase Planes

Mathematics developed:

> analyzing in phase plane the solution of first-order nonlinear ordinary differential equations; concept of equilibria and their stability

6.1 Introduction

We shall consider nonlinear population models in this chapter and introduce a qualitative method for analyzing nonlinear equations.

6.2 Population Models

We consider population models of the form

$$\frac{dN}{dt} = F(N, t), \qquad (6.1)$$

where $F(N, t)$ is the growth rate of a population of, say, humans or fish, of density N.

The simplest population model is given by

$$F = rN,$$

where $r =$ constant is the net growth rate per unit population. This model has been considered in a previous chapter. Its solution leads to an exponential growth in the population:

$$N(t) = N_0 e^{rt}.$$

This behavior is probably realistic only during the initial growth at low population density, when resource constraints on further growth have not yet come into play. Figure 6.1 shows the U.S. census data from

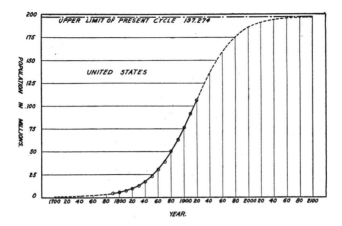

Figure 6.1. Early U.S. census data fitted to a logistic curve.
(From Pearl and Reed [1920].)

the 19th century. The growth of the population in the new continent appeared exponential for about a century after 1750.

The exponential growth stage certainly cannot continue for very long. Sooner or later the population growth taxes the support system (such as food, space, farmland, etc.) and creates overcrowding, which may lead to diseases. A more reasonable model for the growth rate F has the growth slow down as N increases. A commonly used, but controversial, model that has this behavior is the so-called *logistic growth* model, originally proposed by the 19th century Belgian mathematician Pierre François Verhulst.

$$\frac{dN}{dt} = F(N, t) = r N \left(1 - \frac{N}{K} \right), \quad K > 0. \tag{6.2}$$

The solution can be obtained by separation of variables (exercise 3) as:

$$N(t) = \frac{K}{1 + e^{a-rt}},$$

where a is a constant of integration that can be determined by the initial condition

$$N(0) = N_0 = \frac{K}{1 + e^a}.$$

For large t, the solution approaches the asymptotic value of K, i.e.,

$$\lim_{t \to \infty} N(t) = K.$$

This behavior is plotted as a dashed line in Figure 6.1.

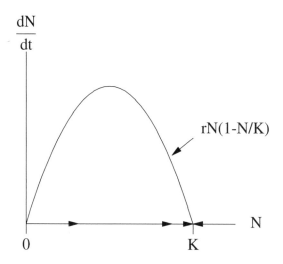

Figure 6.2. F has a parabolic shape and it has
two zeros, at $N^* = 0$ and $N^* = K$.

Verhulst called this solution the "logistic curve." The French term
"logistique" was used to signify the art of calculation. Later, in 1920,
this curve was reintroduced by American scientists Raymond Pearl and
Lowell J. Reed and promoted as a natural "law" for the growth of any
population. Never mind that their fit of U.S. population data in Figure
6.1 covered less than one third of the logistic curve, giving no hint of
the eventual saturation limit. Many other models of population growth
probably would fit the data just as well.

We shall for convenience use the logistic model as one of the examples of nonlinear population models of the form of Eq. (6.1) to illustrate
the use of qualitative methods for their analysis.

For a derivation of this model, see the exercises.

6.3 Qualitative Analysis

Since F is also $\frac{dN}{dt}$, a positive F indicates that N is increasing with time.
We indicate this in a plot of $\frac{dN}{dt}$ vs. N, called the phase diagram (Figure
6.2), by an arrow pointing towards larger N, for those N's that make
$F(N, t)$ positive. At the zeros of $F(N)$, we have $\frac{d}{dt}N = 0$. These are *equilibria* of Eq. (6.2). These equilibria are either *stable* or *unstable*, depending
on whether a small perturbation from the equilibrium solution will lead
it back to or move it away from the equilibrium solution, respectively.
We see from Figure 6.2 that $N^* = K$ is a *stable* equilibrium because
increasing N from N^* will create a negative $\frac{dN}{dt}$, and decreasing N from
N^* will create a positive $\frac{dN}{dt}$. These then tend to move N back to $N^* = K$

(see Figure 6.2). K is called the *carrying capacity* of the population. It can be inferred observationally from the "equilibrium" population, in the absence of external inference. The quantity r can be measured by the net growth rate of a small population, as

$$r \cong \frac{1}{N} \frac{d}{dt} N \quad \text{for } N/K \ll 1.$$

6.4 Harvesting Models

As an exercise involving nonlinear population models we now consider the harvesting of fish. A good reference is the book by Wan (1989). The logistic model of fish growth is used as one of the examples. We assume that the carrying capacity K is the natural population of the fish in a particular fishery in the absence of harvesting. The object of the study is to determine the optimal amount of fish one can harvest from a fishery in a sustainable manner.

Let H be the harvest rate. We consider, then,

$$\frac{d}{dt} N = r N \left(1 - \frac{N}{K} \right) - H. \tag{6.3}$$

The harvest rate in an unregulated fishery should be proportional to N: the higher the fish population, the higher the catch. The proportionality constant should be dependent on E, the level of effort spent to fish. Typically

$$H = qEN, \tag{6.4}$$

with q being a proportionality constant. qE is known as the "fishing mortality"; it has the same dimension as r.

The new equation, now incorporating logistic growth and harvesting,

$$\frac{d}{dt} N = r N \left(1 - \frac{N}{K} \right) - qEN, \tag{6.5}$$

has two equilibria. These can be found by setting $\frac{d}{dt} N = 0$ in Eq. (6.5). They are

$$N_1^* = 0 \quad \text{and} \quad N_2^* = K(1 - qE/r). \tag{6.6}$$

The equilibrium N_2^* is stable, as can be inferred from Figure 6.3, which shows that perturbations away from N_2^* will tend back to N_2^*.

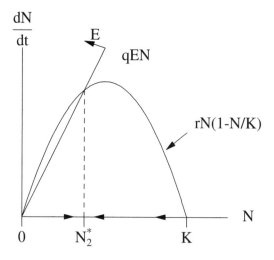

Figure 6.3. This diagram plots each of the two terms on the right-hand side of Eq. (6.5) against N. $\frac{dN}{dt}$ is the difference of the two curves plotted. $N^* = N_2^*$ is a stable equilibrium.

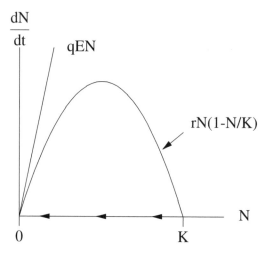

Figure 6.4. Now, $N_1^* = 0$ is stable. Extinction of the fish stock is the only stable solution.

This equilibrium population, although less than the natural carrying capacity K, is sustainable. That is, it is sustainable as long as $qE/r < 1$ in Eq. (6.6). When the fishing mortality qE is greater than the growth rate r ($qE/r > 1$), N_2^* becomes negative and we have only one equilibrium ($N^* = 0$). The stability behavior of the system also changes (see Figure 6.4).

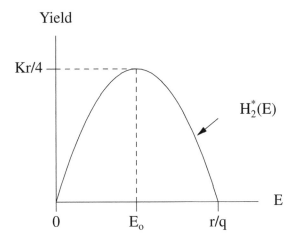

Figure 6.5. Sustained yield curve as a function of fishing effort E. The optimal effort is given by $E = E_0 \equiv r/2q$. The maximum yield is $H = \frac{1}{4}Kr$.

To have a sustainable harvest, one must have

$$qE/r < 1.$$

The corresponding harvest rate, called the *sustained yield*, is given by

$$H(N_2^*, E) = qE N_2^* = qEK(1 - qE/r) \equiv H_2^*.$$

The yield corresponding to the other equilibrium, $N_1^* = 0$, is obviously zero:

$$H(N_1^*, E) = 0 \equiv H_1^*.$$

The sustained-yield curves, $H_2^*(E)$, are shown in Figure 6.5, as a function of E.

6.5 Economic Considerations

Economic considerations would often lead to a nonoptimal harvesting effort. Let the price per unit fish be p and the cost per unit effort be c, both assumed to be constants. Then the total cost (per unit time) is cE, while the revenue from the harvest (per unit time) is pH. In a sustained-yield situation, the break-even point would be the intersection of $pH_2^*(E)$ and cE.

The break-even point occurs (when $H_2^*(E) = (c/p)E$) at $E = E_{be} = \frac{r}{q} \cdot (1 - \frac{(c/p)}{qK})$. See Figure 6.6.

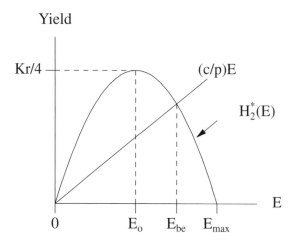

Figure 6.6. $E_{max} = r/q$ is the maximum effort beyond which the fish population will become extinct. E_{be} is the break-even effort for a given cost c and price p.

For higher priced fishes, this break-even effort occurs for an E closer and closer to E_{max}. But since the yield for these values of E is getting closer and closer to zero, the effort E will never get to E_{max} unless $c/p \to 0$. So, for a fish whose price is finite (i.e., $c/p > 0$), E_{be} is less than E_{max} by some finite amount.

If the current level of effort expended is $E < E_{be}$, i.e., less than the break-even level, then there would still be profit to be made for additional effort. There is, however, no economic incentive to expend an effort greater than the break-even point E_{be}. Since, as we have pointed out earlier, E_{be} is less than E_{max} by a finite amount, the fish will not go extinct.

6.6 Depensation Growth Models

The above conclusion does not seem to be consistent with our experiences: along the Pacific coast there were collapses of the sardine fishery near San Francisco and the anchovy fishery in Peru. This forces us to re-examine our model.

It is known that clupeids (such as sardines, anchovies, and herrings) form large and closely packed schools. Natural predators, which are not similarly concentrated, may be less efficient at decimating the prey population. On the other hand, if their population density gets too low these prey species are unable to reproduce rapidly enough to compensate for predator mortality. These populations grow only

F(N) or dN/dt

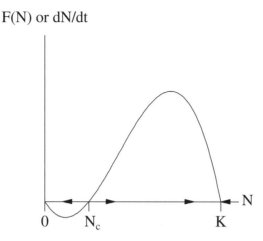

Figure 6.7. Depensation growth model without harvesting.

when their N lies between a (critical mass) value, N_c, and the carrying capacity, K. An example of such "depensation" growth models is

$$F(N) = r N \cdot (N/N_c - 1) \cdot (1 - N/K). \qquad (6.7)$$

There are now three equilibrium solutions (see Figure 6.7), with $N = 0$ being a *stable* equilibrium (compare this situation with the earlier logistic model, where $N = 0$ is an unstable equilibrium). If the population falls below its critical mass value, N_c, it will head to $N = 0$. Extinction is a stable equilibrium.

With the harvest term put back in, we now have

$$\frac{d}{dt} N = -r \cdot N \cdot (1 - N/N_c) \cdot (1 - N/K) - qEN. \qquad (6.8)$$

There are three equilibria: $N_1^* = 0$ and N_3^* are stable, and N_2^* is unstable. These equilibria are obtained by setting $dN/dt = 0$ in Eq. (6.8), and solving

$$N^* \cdot [r(N^*/N_c - 1)(1 - N^*/K) - qE] = 0.$$

Alternatively, they can be obtained as points of intersection of the harvesting line, qEN, and the growth rate, $F(N)$, in Figure 6.8.

Without regulatory control, a catastrophic collapse of the fishery or an outright extinction of the species is a distinct possibility for populations with a depensation growth behavior (see Figures 6.9 and 6.10).

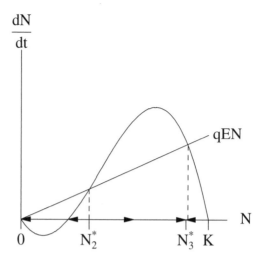

Figure 6.8. Depensation growth model with harvesting.

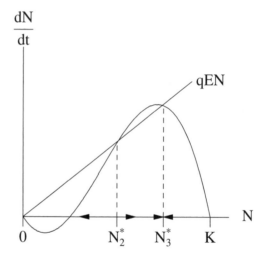

Figure 6.9. At higher levels of effort E, the unstable equilibrium N_2^* and the stable equilibrium N_3^* come closer to each other.

To see this in more detail, consider the sustained yield,

$$H_3^* \equiv H(N_3^*) = qE N_3^*,$$

as a function of E (see Figure 6.11). (Remember that N_3^* is a function of E also.) Let $E = E_{\max}$ be the largest effort beyond which the fish population will head for extinction. In this case, this occurs when

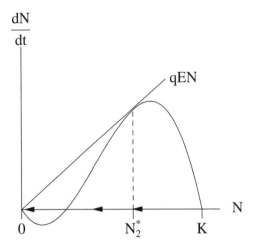

Figure 6.10. Until $N_2^* = N_3^*$. Then suddenly the stability properties of the equilibria change. N_3^* becomes *unstable*. All solutions will head to the only stable equilibrium, $N_1^* = 0$.

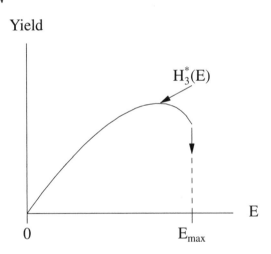

Figure 6.11. Sustained yield curve for the depensation growth model.

$N_3^* = N_2^*$, giving, as can be shown,

$$E_{\max} = \frac{r}{q}\left[\frac{\left(\frac{K+N_c}{2}\right)^2}{N_c \cdot K} - 1\right].$$

At this level of harvesting, the yield is still quite high (in contrast to the previous model, where the yield is zero at $E_{\max}$). A fishery producing a high yield (and possibly at a comfortable profit) may, without knowing it, allow a harvesting effort exceeding E_{max}. Suddenly the yield is not sustainable and the fish population declines monotonically to zero.

6.7 Comments

Myers et al. (1995) analyzed data from 128 fish stocks in order to determine if depensatory dynamics are common among exploited fish populations. Of the 128 fish stocks examined, they found statistically significant depensation only in an Icelandic herring stock and two Pacific salmon stocks. In all cases there was evidence of environmental effects that probably underlie the reduced rate of population increase. (On the other hand, critics point to the fact that 80% of the 128 fish stocks considered by Myers et al. yield inconclusive results because of a lack of statistical power.)

6.8 Exercises

1. *Regulated harvesting*

Study Eq. (6.3) for a constant harvesting rate H in two ways:

a. Use phase planes to find graphically the equilibria and their stability.

b. Solve Eq. (6.3) exactly using the method of separation of variables.

2. *SI model of epidemiology*

Derive a differential equation modeling the spread of flu among individuals in a fixed population of N. That is, express the rate of increase of the number of infected individuals ($I(t)$) in terms of $I(t)$:

$$\frac{dI}{dt} = \cdots .$$

You can assume that (a) the disease is nonfatal, and so N is constant; (b) the number of susceptible individuals $S(t)$ is the same as the number of uninfected individuals at time t; and (c) the rate of increase of infected individuals is proportional to the number of infected times the number of susceptible individuals at any given time. Obtain a single equation for $I(t)$ by eliminating $S(t)$.

3.   Solve via separation of variables the following logistic equation for $I(t)$

$$\frac{dI}{dt} = rI \cdot \left(1 - \frac{I}{K}\right), \quad I(0) = I_0,$$

where r and K are positive constants. Find also the limit of $I(t)$ as $t \to \infty$.

4.   Obtain the explicit solution to the harvesting problem

$$\frac{dN}{dt} = rN(1 - N/K) - H(N), \quad H = qEN,$$

subject to the initial condition $N(0) = N_0$. Here $N = N(t)$. The parameters r, K, q, and E are positive constants. Deduce the two limiting behaviors as $t \to \infty$.

5.   There are many important classes of species whose birthrate is not proportional to the population size $N(t)$. Suppose, for example, that each member of the population requires a partner for reproduction and that each member relies on chance encounters to meet a mate. If the expected number of encounters is proportional to the product of the numbers of males and females, and if these are equally distributed in the population, then the number of encounters, and hence the birthrate, is proportional to $N(t)^2$. The death rate is still proportional to $N(t)$. Consequently, the population size $N(t)$ satisfies the differential equation

$$\frac{dN}{dt} = bN^2 - aN;$$

a, b are positive constants.

a. Solve for $N(t)$ given $N(0)$.

b. Find the long-time behavior by taking $t \to \infty$ in your solution in (a). Do this for the case $N(0) < a/b$.

6. Right-handed snails

D'Arcy Wentworth Thompson, a noted scientist of natural history, wrote in his book, *On Growth and Form* (1917): "But why, in the general run of shells, all the world over, in the past and in the present, one direction of twist is so overwhelmingly commoner than the other, no man knows." Most snail species are *dextral* (right-handed) in their shell pattern. *Sinistral* (left-handed) snails are exceedingly rare.

A plausible model for the appearance of such a bias in population handedness can be as follows: let $p(t)$ be the ratio of dextral snails in the population of snails. $p = 1$ means that all snails are right-handed, and $p = 0$ means that all snails are left-handed. A model equation for $p(t)$ can be

$$\frac{d}{dt}p = \alpha p \cdot (1 - p) \cdot \left(p - \frac{1}{2}\right),$$

which has no left–right bias.

a. Locate the equilibria of p and determine their stability.

b. Suppose that at $t = 0$ (which is a very long time ago, perhaps a few hundred million years ago), $p(0) = \frac{1}{2}$; that is, the dextral and sinistral snails are evenly divided. Describe what will happen a few hundred million years later. Argue that we should not expect that $p(t) = \frac{1}{2}$ as $t \to \infty$ (i.e., an equal number of dextral and sinistral snails at the present time), and argue that our present state of affairs (mostly dextral snails) is an accident (i.e., we could just as well have mostly sinistral snails now).

7. Firefly synchrony

Philip Laurent wrote in *Science* in 1917 about a phenomenon he saw in Southeast Asia: "Some twenty years ago I saw, or thought I saw, a synchronal or simultaneous flashing of fireflies. I could hardly believe my eyes, for such a thing to occur among insects is certainly contrary to all natural laws."

Joy Adamson wrote in 1961 about an African version of the same phenomenon: "A great belt of light, some ten feet wide, formed by thousands upon thousands of fireflies whose green phosphorescence bridges the shoulder-high grass . . . One is left wondering what means of communication they possess which enables them to coordinate their shining as though controlled by a mechanical device."

It is the males that flash to attract females. Individual males have their own light-emitting oscillators, with a natural period of about 0.9 second, but also have some ability to change the period to mimic the flashing of a particularly strong neighbor, thus increasing their own attractiveness to the females. Some species of fireflies can do this better than others, being able to change the frequency of their flashing by up to 15% in response to an external stimulus.

Let $\theta_e(t)$ be the phase (of the angle in radians) in the flashing of the external stimulus at time t (e.g., an attractive male neighbor). The frequency ω_e of the flashing is assumed to be given and fixed with

$$\frac{d\theta_e}{dt} = \omega_e.$$

Let $\theta(t)$ be the phase of a particular firefly under consideration at the same time t. If $\theta(t)$ is ahead of $\theta_e(t)$, it will try to slow down. If $\theta(t)$ is behind, it will try to speed up. So a model for $\frac{d}{dt}\theta$ is given by:

$$\frac{d\theta}{dt} = \omega + \alpha \sin(\theta_e - \theta).$$

The size of the external stimulus is measured by the parameter α, and ω is the natural frequency of this firefly in the absence of an external stimulus. So if θ_e is slightly ahead of θ, θ will increase so that $\frac{d}{dt}\theta > \omega$. It will slow down ($\frac{d}{dt}\theta < \omega$) if θ_e is slightly behind θ. Let $\phi(t) = \theta_e - \theta$ be the relative phase, $\tau = \alpha t$, and $\delta = (\omega_e - \omega)/\alpha$. The above equation becomes

$$\frac{d\phi}{d\tau} = \delta - \sin\phi,$$

involving only one parameter, δ. δ can be positive or negative; we consider the positive case below. The negative case is similar.

a. Plot the phase diagram (of $\frac{d\phi}{d\tau}$ vs. ϕ) schematically for the three different cases ($\delta = 0, 0 < \delta \leq 1, \delta > 1$) and for the range $-\pi \leq \phi \leq \pi$.

b. Describe the equilibria in the three cases and their stability (using arrows to indicate the direction of increasing ϕ).

c. Describe what happens to $\theta(t)$ eventually in the three cases: (i) for the very strong stimulus case, $\delta = 0$; (ii) for the weak stimulus case, $\delta > 1$; and (iii) for the moderate and realistic case, $0 < \delta \leq 1$. We call the case of nonsynchronous change in phase a phase drift, and we call the establishment of a steady phase relative to the external phase a phase locking. When phase locking occurs, the fireflies flash at the same frequency.

d. Derive the condition (an inequality among ω, α, and ω_e) for phase locking.

8. Flu epidemic

A flu epidemic starts in an isolated population of 1,000 individuals. When 100 individuals are infected the rate at which new infections are occurring is 90 individuals per day. If 20 individuals are infected at time $t = 0$, when will 90% of the population be infected?

(*Hints and assumptions*: Let $I(t)$, $N(t)$ be the number of infected and noninfected individuals at time t, respectively. The number of noninfected individuals is $N(t) = 1000 - I(t)$. The rate at which $I(t)$ is increasing should be proportional, $I(t)N(t)$.)

9. *Modeling the spread of technology*

Let N^* be the total number of ranchers in Uruguay, and $N(t)$ be the number of ranchers who have adopted an improved pasture technology there. Assume that the rate of adoption, $\frac{dN}{dt}$, is proportional to both the number who have adopted the technology and the fraction of the ranchers who have not (and so are susceptible to conversion). Let a be the proportionality constant.

a. Write down the differential equation for $N(t)$.

b. According to Banks (1993), $N^* = 17{,}015$, $N(0) = 141$, $a = 0.490$ per year. Determine how long it takes for the improved pasture technology to spread to 80% of the population of the ranchers.

10.   For a fish population modeled by a depensation growth model, we have, when there is harvesting,

$$\frac{d}{dt} N = F(N) - H(N),$$

where

$$F(N) = r\, N \cdot (N/N_c - 1)(1 - N/K)$$

and $H(N) = qEN$.

a. Find (and plot, freehand) the sustained yield $H(N_3^*)$ as a function of effort E, and the unsustainable yield $H(N_2^*)$, also as a function of E (and plot on the same figure). Here N_3^* is the nontrivial stable equilibrium of N and N_2^* is the unstable equilibrium.

b. Find the $E(= E_{\max})$, where the two curves in (a) merge (this happens when $N_2^* = N_3^*$). What happens to the fishery and the fish population when $E = E_{\max}$?

c. Suppose harvesting is done at effort level $E = E_{\max}$ for a while so that the fish population is below what was thought of as the "sustainable" N_3^* (which is equal to N_2^*, so actually it is not really sustainable). Realizing that there is a problem, the government puts in a fishing limit that reduces the effort E to slightly below $E_{\max}$. Can the fishery recover?

7

Discrete Time Logistic Map, Periodic and Chaotic Solutions

Mathematics introduced:
> overshoot instability; nonlinear map; linear and nonlinear stability; periodic and aperiodic solutions; deterministic chaos; sensitivity to initial conditions

7.1 Introduction

For population dynamics governed by a continuous first-order differential equation of the form studied in the previous chapter,

$$\frac{d}{dt}N = F(N),$$

the behavior of the solution is understood qualitatively by looking at the equilibria (also called the *fixed points*) of the equation. The long-term behavior of the solution is given by a monotonic approach to the stable equilibrium. There are no periodic oscillations and there is no erratic meandering. Initial conditions are unimportant in this picture.

Yet a richer time-dependent behavior is often seen in animal populations, including apparently random fluctuations in the size of their colonies. How should the simple models be modified so as to yield a richer, and more realistic, set of possible behaviors?

There are two approaches. One is to regard the fluctuations as being forced by external factors, such as by population immigration or environmental changes that affect the carrying capacity of the population. For example, if we are studying the logistic equation as a simple model for the population, i.e.,

$$\frac{d}{dt}N = rN\left(1 - \frac{N}{K}\right),$$

we may want to specify some fluctuations in the carrying capacity $K(t)$, or we may add an "immigration" term, $I(t)$, to the right-hand side of the above equation. If these externally specified terms have random fluctuations, so will the population $N(t)$ they induce.

Another approach is to inquire if the more interesting time behavior, including the seemingly random cases, can arise from a simple deterministic equation like this without needing outside influence. Robert May's (1974) paper showed that simple population dynamics can produce intrinsically interesting periodic or erratic solutions that are, on the surface, indistinguishable from those induced by external factors. There is an account of the discovery's impact on thinking in the field of theoretical ecology in the 1970s in the book *It Must Be Beautiful*, edited by Graham Farmelo.

Logistic Growth for Nonoverlapping Generations

May (1974) pointed out that while in some biological populations (such as humans) growth is a continuous process, generations overlap, and the appropriate mathematical description involves nonlinear differential equations, in other biological situations (such as 13-year periodical cicadas), population growth takes place at discrete time intervals and generations do not overlap. In these latter cases, nonlinear *difference* equations should be the more appropriate mathematical description.

Robert May considered the discrete version of the logistic equation

$$\frac{N((n+1)\Delta t) - N(n\Delta t)}{\Delta t} = r N(n\Delta t) \left[1 - \frac{N(n\Delta t)}{K} \right]. \tag{7.1}$$

It is more appropriate for populations in which births occur in well-defined breeding seasons, and the generations (Δt) are nonoverlapping, than the continuous version:

$$\frac{dN}{dt} = r N \left(1 - \frac{N}{K} \right), \tag{7.2}$$

which assumes that births are occurring continuously.

Equilibrium solutions are located at $N_1^* = 0$, $N_2^* = K$ for both discrete and continuous versions. (See Figure 7.1.)

For the continuous system, the stability of each of these two equilibria can be inferred easily from a plot of dN/dt vs. N. As can be seen in the left panel of Figure 7.1, the equilibrium $N^* = 0$ is unstable, while $N^* = K$ is stable, judging by the arrows, which show the direction of change of N. On the right panel of Figure 7.1, we make a similar plot of the rate of change of N in one generation: $(N(t + \Delta t) - N(t))/\Delta t$, vs. N. Again, the direction of the arrows indicates the direction of change of N. We see that the equilibrium $N^* = 0$ is unstable. The equilibrium $N^* = K$

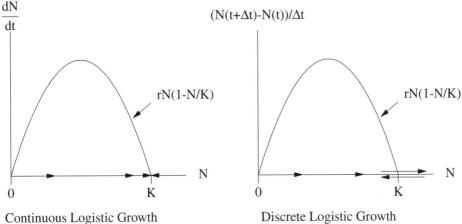

Figure 7.1. Continuous logistic growth (left) vs. discrete logistic growth (right). For the discrete version, there is the possibility that the change in population $N(t)$ in one generation (Δt) can be so large that it overshoots the equilibrium point (K), leading to an oscillatory instability.

is usually stable (as arrows point toward it), but an interesting phenomenon of overshoot may happen for large growths in one generation ($r \Delta t > 2$; see below). That is, if the growth rate is high, and the length of a generation is long, the change of the population from one generation to the next may be so large that N "overshoots" the equilibrium, with an ensuing oscillatory instability as a result of repeated back and forth overshoot. This instability is new and is not present in the continuous system. This new instability usually gives rise to periodic solutions, with the period depending on intrinsic parameters in the model without any external influence. The behavior of a system without even a single stable equilibrium present is more complicated but interesting.

7.2 Discrete Map

The difference equation (7.1) can be rewritten in the form of a mapping:

$$N_{n+1} = f(N_n), \tag{7.3}$$

where

$$f(N_n) = N_n + r \Delta t N_n \left(1 - \frac{N_n}{K}\right) \tag{7.4}$$

and $N_n \equiv N(n\Delta t)$.

The solution can be obtained by applying the mapping successively. That is, starting with a given N_0, Eq. (7.3) gives us N_1. Knowing N_1, it then yields N_2, etc.

In this formulation, the equilibrium N^* is determined from solving

$$N^* = f(N^*). \tag{7.5}$$

This equilibrium is stable to small perturbations if

$$|f'(N^*)| < 1. \tag{7.6}$$

This is because if we write

$$N_n = N^* + u_n$$

and assume u_n is small, Eq. (7.3) becomes approximately

$$u_{n+1} = f'(N^*)u_n.$$

Thus $f'(N^*)$ is the "amplifying factor" from u_n to u_{n+1}. For stability its magnitude should be less than 1.

There are two equilibria, satisfying Eq. (7.5):

$$N^* = 0 \text{ and } N^* = K.$$

Since

$$f'(N^*) = 1 + r\Delta t - 2\frac{r\Delta t}{K}N^*,$$

$$f'(0) = 1 + r\Delta t, \quad f'(K) = 1 - r\Delta t,$$

the equilibrium $N^* = 0$ is unstable. The equilibrium $N^* = K$ is usually stable, unless $r\Delta t > 2$. The three different behaviors of N_{n+1} vs. N_n near $N^* = K$ are shown in Figure 7.2 for different cases of $f'(K)$. For $0 > f'(K) > -1$, the solution is asymptotically stable as each iteration brings the solution closer and closer to the equilibrium. (From another perspective, the perturbation u_n from the equilibrium becomes smaller and smaller as n increases.) For $f'(K) = -1$, the solution approaches a distance from N^*. From there the perturbation neither grows nor decays. Instead the perturbation executes a periodic oscillation around the equilibrium, changing sign after each iteration but repeating after two iterations (i.e., $u_{n+2} = f'(K)u_{n+1} = f'(K)f'(K)u_n = u_n$). This is

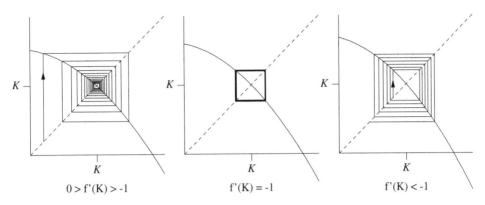

Figure 7.2. Solution of the discrete logistic equation near the equilibrium point. The horizontal axis is N_n and the vertical axis is N_{n+1}. The curve is $f(N_n) = N_n + r N_n(1 - N_n/K)$. The dashed line is $N_{n+1} = N_n$. Three cases of linear stability are shown: asymptotically stable, borderline stable (periodic oscillatory), and unstable.

called a 2-cycle; it is periodic with period $2\Delta t$. For $f'(K) < -1$, small perturbations grow. The solution is therefore linearly unstable. Nonlinearly it is not necessarily unstable, as we shall see.

7.3 Nonlinear Solution

We next solve the discrete logistic equation graphically for various values of the parameter $r \Delta t$. We construct a plot (see Figure 7.3) whose horizontal axis is N_n and whose vertical axis is N_{n+1}. On this plot we first draw $f(N_n)$ as a function of N_n. The graph is in the form of a parabola. Next we draw a straight line $N_{n+1} = N_n$ (a 45° line from the origin). The intersection of this straight line with the parabola then yields the equilibrium solution N^*.

The time-dependent solution to Eq. (7.1) can also be constructed from this plot. Given an N_0, we locate its value on the horizontal axis. We move up vertically until we hit the parabola, $f(N_0)$. This is then the value N_1 (from Eq. (7.3)), which we read off the vertical axis. Next we want to locate this N_1 on the horizontal axis; this is facilitated by the 45° line. That is, we use N_1 as the starting point on the vertical axis, we move horizontally until we hit the 45° line, and then we move down vertically until we hit the horizontal axis. This is the desired location for N_1.

We repeat this process to find N_2, N_3, etc. The resulting graph looks like a cobweb, and thus it is called a cobweb map. The graphical process we outlined above is called "cobwebbing."

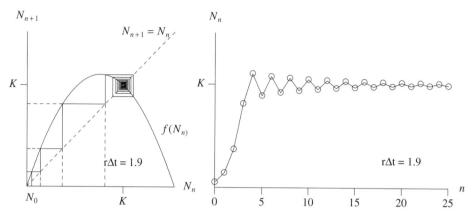

Figure 7.3. Graphical construction of the solution of the discrete logistic equation for the linearly stable case of $r \Delta t = 1.9$ (left panel). The curve is $f(N_n) = N_n + r N_n(1 - N_n/K)$. The diagonal dashed line is $N_{n+1} = N_n$. The right panel shows the solution $N(t)$ as a function of discrete $t = n\Delta t$. (Modified from original figure in Kot [2001], by permission.)

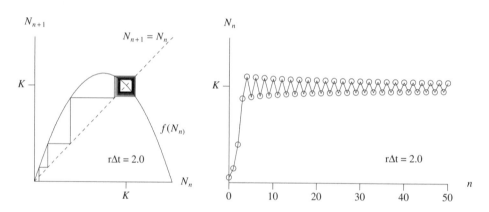

Figure 7.4. Same as Figure 7.3 except for the borderline case of $r \Delta t = 2.0$. Nonlinearly the solution remains the same as the linear solution: a 2-cycle periodic solution about (and close to) the equilibrium.

Figure 7.3 shows the cobwebbing construction and the solution to Eq. (7.3) for the stable case ($r \Delta t = 1.9$). The solution approaches asymptotically to the equilibrium $N^* \equiv K$. So the nonlinear stability of this equilibrium is the same as the linear determination.

Figure 7.4 is for the case of $r \Delta t = 2.0$. The nonlinear evolution is similar to the linear one. The solution is a 2-cycle.

Figure 7.5 is for the case of $r \Delta t = 2.2$. For this case, small perturbations from $N^* = K$ grow and therefore the equilibrium is linearly

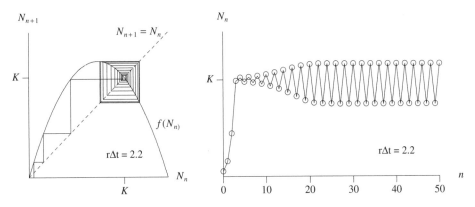

Figure 7.5. Same as Figure 7.3 except for the linearly unstable case of $r\Delta t = 2.2$. Nonlinearly the solution increasingly deviates from the equilibrium point until it settles down to a nonlinear 2-cycle periodic solution. (Modified from original figure in Kot [2001], by permission.)

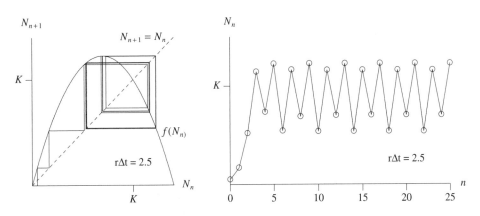

Figure 7.6. Same as Figure 7.3 except for $r\Delta t = 2.5$. The solution is a 4-cycle. (Modified from original figure in Kot [2001], by permission.)

unstable. However, our exact solution (by this graphical method) shows that the solution is nonlinearly periodic, with period $2\Delta t$. That is, the solution after a while repeats itself every two iterations.

Figure 7.6 is for the case of $r\Delta t = 2.5$. A 4-cycle periodic solution is found.

An 8-cycle solution is found for $r\Delta t = 2.55$ (see Figure 7.7).

There are more and more period doublings as $r\Delta t$ increases until at or above $r\Delta t = 2.5699456\ldots$ a period 3 appears. Then all periods are present. The solution does not approach N^* but is nevertheless

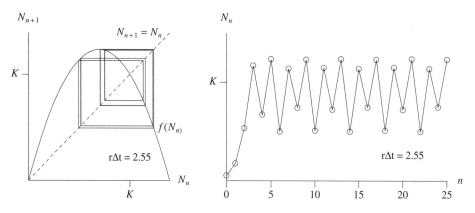

Figure 7.7. Same as Figure 7.3 except for $r\,\Delta t = 2.55$. The solution is an 8-cycle. (Modified from original figure in Kot [2001], by permission.)

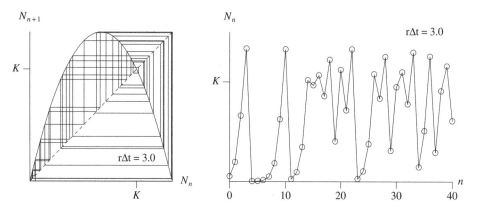

Figure 7.8. Same as Figure 7.3 except for $r\,\Delta t = 3.0$. The solution is aperiodic and bounded (chaos). (Modified from original figure in Kot [2001], by permission.)

bounded. It never repeats itself, and so is called *aperiodic*. Professor James Yorke at the University of Maryland coined the term "chaos" to describe this behavior in his paper with his student Tien-Yien Li entitled "Period Three Implies Chaos" (Yorke and Li, 1975).

The behavior for $r\,\Delta t = 3.0$ in Figure 7.8 is typical of this aperiodic behavior.

7.4 Sensitivity to Initial Conditions

A *chaotic* time series is one that is aperiodic. In addition, a chaotic solution is extremely sensitive to the initial condition.

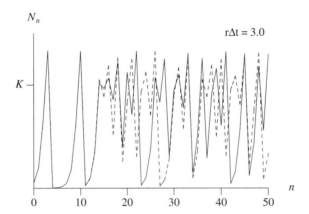

Figure 7.9. The dashed curve is the solution under the same condition as the solid curve, except with a slightly different initial condition. (Modified from original figure in Kot [2001], by permission.)

In Figure 7.9, the solution from Figure 7.8 is shown again in the solid line. The dashed line shows the solution starting with a slightly different initial condition ($N_0/K = 0.050001$ instead of 0.05). They are indistinguishable from each other for about 15 iterations. Then they diverge from each other dramatically. Loss of predictability is typical of chaotic systems.

7.5 Order Out of Chaos

Suppose we are given a time series of a variable (e.g., the left panel of Figure 7.10) and are told that it is all right to assume that there is only one degree of freedom. The behavior of this time series appears chaotic. We want to know if there is some underlying order to it. In particular, we want to find the equation (the map) that governs its evolution.

Looking back at earlier pages of this lecture we see that for $N_{n+1} = f(N_n)$, the functional form of f can be deduced by plotting the pairs (N_0, N_1), (N_1, N_2), (N_2, N_3), ..., etc. in an N_n, N_{n+1} plot. These points should lie on a more regular shape, a parabola in our previous example (see the right panel of Figure 7.10).

So, by plotting this way, we are discovering the form of $f(N)$ (the mapping) and hence the original deterministic equation that governs the apparently disorganized data. That is, we are uncovering the underlying order. Of course this will not work so easily if the chaotic time series is the result of a higher dimensional mapping.

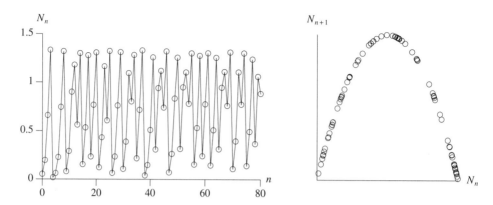

Figure 7.10. The left panel is a time series, which in this case appears random. The right panel is constructed by plotting the pair (N_n, N_{n+1}) from the data in the left panel.

7.6 Chaos Is Not Random

Before chaos was understood as an aperiodic solution to deterministic equations, with the property of hypersensitivity to initial conditions, chaotic behavior like that depicted in the left panel of Figure 7.10 was often thought of as random. In fact, in the late 1940s, the famous Hungarian-American mathematician and father of game theory, John von Neumann, even suggested using the logistic equation with $r \Delta t = 3$ as a random number generator (actually, $x_{n+1} = 4x_n(1 - x_n)$). Had the casinos adopted it, we would have been able to figure out, using the right panel of Figure 7.10, what the next outcome of a slot machine would be given its current state!

7.7 Exercises

1. Periodic solutions and their loss of stability

The logistic difference equation

$$(N_{n+1} - N_n)/\Delta t = r \cdot N_n \cdot (1 - N_n/K)$$

possesses chaotic solutions for $r \Delta t > 2.569946\ldots$. The above equation can be made dimensionless and rewritten in the simpler form:

$$X_{n+1} = f(X_n), \tag{7.7}$$

where

$$f(x) = ax(1 - x), \quad a \equiv 1 + r\Delta t, \quad a > 1$$

and

$$X_n \equiv N_n \cdot r\Delta t/((1 + r\Delta t)K).$$

We want to investigate the equilibrium solutions to this equation and their loss of stability as a increases.

a. Find the equilibria of Eq. (7.7) and determine their stability.

b. For $a > 3$, the solution to Eq. (7.7) includes a 2-cycle. That is, it obeys, along with the equilibria you just found in (a), this new equation:

$$X_{n+2} = X_n = X_2.$$

We know

$$X_{n+2} = f(X_{n+1}) = f(f(X_n)) \equiv g(X_n). \tag{7.8}$$

Find $g(x)$, and then solve for the equilibrium solution of Eq. (7.8), which satisfies $X^* = g(X^*)$. You should recover the equilibria in (a), plus two new ones if $a > 3$. These two new equilibria are not real for $a < 3$. In trying to solve for the equilibria you may need to solve a quartic equation, but you can take advantage of the fact that you already know two of the four solutions. After you factor these two factors out you are left with a quadratic equation to solve. To save you some algebra, it is factored for you here:

$$X^*[aX^* - (a - 1)][a^2 X^{*2} - a(a + 1)X^* + (a + 1)] = 0.$$

c. What do you think is the nature of the two new equilibrium solutions you have found in (b)? (That is, are they true steady state solutions? You don't need to know their value to decide this. Just use the fact that they are the equilibria of (7.8) but not (7.7).)

d. Describe how you would go about determining the stability of the equilibria found in (b) (but do not actually do the stability calculation for each equilibrium; it is quite messy). Although you are not asked to do this, you may be interested to know that the 2-cycle solutions lose their stability when $a > 1 + \sqrt{6}$. Then 4-cycle solutions emerge, which can be obtained from

$X_{n+4} = f(X_{n+3}) = f(f(f(f(X_n))))$. The process can be repeated for n-cycle solutions as a increases further.

2. Beverton–Holt model

A goal of fishery scientists has been to determine the relationship between the size of a fish population and the number of offspring from it that survive to enter the fishery. Fishery scientists refer to a population of a fish as a stock, and the process of entering the fishery as recruitment. The relationship, called the stock-recruitment relationship, is often obtained by empirical fitting of the data to an analytic function. The Beverton–Holt stock recruitment is governed by the difference equation

$$N_{n+1} = \frac{RN_n}{1 + [(R-1)/K]N_n}, \quad R > 1.$$

a. Find the equilibria and determine their stability.

b. Find an exact, closed form solution. (*Hint*: Use the substitution $X_n = 1/N_n$. The equation for X_n turns out to be linear.) This is one of the rare cases where a nonlinear difference equation can be solved exactly.

3. Ricker model

Consider the following density-dependent difference equation for population density N_n at time n:

$$N_{n+1} = N_n e^{r[1-(N_n/K)]} \equiv f(N_n),$$

where $r > 0$ is the growth rate and $K > 0$ is the carrying capacity. This model is preferred by some over the discrete logistic model because N_t is always positive if N_0 is positive.

a. Find the equilibria N^* and discuss their linear stability as a function of r. Show that there is a change in stability at $r = 2$ ($r = 2$ is therefore called a bifurcation point).

b. Find the 2-cycle equilibria for $r > 2$ from

$$N_{n+2} = f(f(N_n)) \equiv g(N_n)$$

(i.e., find N^* such that $N^* = f(f(N^*))$). How would you interpret the nature of the two new "equilibria" you find here (which are not the equilibrium of $N_{n+1} = f(N_n)$)?

c. Show that, for large n, the maximum N_n is $f(K/r)$ and the minimum N_n is $f(f(K/r))$.

d. Use cobwebbing or another method to show that the solution becomes chaotic (bounded and aperiodic), for large r.

8

Snowball Earth and Global Warming

Mathematics required:
> Taylor series expansion, or tangent approximation; solution to nonhomogeneous ordinary differential equation

Mathematics developed:
> multiple equilibrium branches, linear stability, slope-stability theorem

8.1 Introduction

In her popular 2003 book, *Snowball Earth, The Story of a Maverick Scientist and His Theory of the Global Catastrophe That Spawned Life as We Know It*, writer Gabrielle Walker followed the adventures of field geologist Paul Hoffman of Harvard University as he pieced together the evidence supporting a dramatic theory: about 600 million years ago our earth was entirely covered by ice. The evidence was laid out by Hoffman and his colleagues in their 1998 paper in *Science* and further described in a *Scientific American* article in 2000 by Hoffman and Schrag. This scenario of global glaciation (see Figure 8.1) (in particular over the equator) was first proposed by Brian Harland of Cambridge University, who called it the Great Glaciation (read the *Scientific American* article by Harland and Rudwick [1964]). The term "snowball earth" was actually coined in 1992 by a Caltech geologist, Joseph Kirschvink. That such a scenario is possible and inevitable was predicted in the 1960s by the Russian climate theorist Mikhail Budyko, using a simple climate model that now bears his name. According to Budyko's model, however, once the earth was completely covered by ice, glaciation would be irreversible because that equilibrium state is stable. Life would die off on land and in the oceans because sunlight could not reach across the thick ice sheets. (We now know that some life-forms could survive even under these harsh conditions, perhaps near hydrothermal vents at the bottom of deep oceans, deriving the energy necessary for life from the heat escaping from earth's molten core. The same geothermal heat would also prevent the ocean from freezing solid.) Hoffman's evidence shows furthermore that there was a dramatic end to the snowball earth. The abrupt end

Figure 8.1. Snowball earth. (Used by permission of W. R. Peltier.)

of the last episode, in an inferno with torrential acid rain, actually led to an explosion of diversity of multicellular life-forms, called the "Cambrian explosion." Before the Cambrian period, the earth had been inhabited by single-cell slimes for over a billion years. We will first present Budyko's model and then discuss some later theories that may explain how the earth deglaciated.

We currently live in a rare period of warm climate over most of the globe, although there are still permanent glaciers over Antarctica, Greenland, Northern Canada, and Russia. Over millions of years in the earth's history, massive ice sheets repeatedly advanced over continental land masses from polar to temperate latitudes and then retreated. We are actually still in the midst of a long cooling period that started three million years ago, punctuated by short *interglacial* periods of warmth lasting about 20,000 years. We are in, and probably near the end of, one of these interglacial respites. *Ice ages* were a norm rather than an exception, at least during much of the time our species evolved into modern humans. The harsh conditions might have played a role in the evolution of our brain as humans struggled to survive in the cold climate.

If we look further back into the paleoclimate record, we find some notable long periods of *equable* climate, when the planet was ice-free and warm conditions prevailed over the globe. (The word *equable* means *even*, and refers to the lack of temperature contrast between the equator and the pole during this period.) The Eocene, some 50 million years ago, is one such period. An earlier one is the Cretaceous, some 65 to 140 million years ago. During the Eocene, alligators and flying lemurs were found on Ellesmere Island (at paleo-latitude 78°N); tropical plants, which are intolerant of even episodic frost, were thriving on

Spitsbergen (at paleo-latitude 79°N); and trees grew in both the Arctic and the Antarctic. During the Cretaceous, palm trees grew in the interior of the Asian continent near paleo-latitude 60°N. We will not discuss models of equable climate much here; you may want to read the paper by Brian Farrell (1990).

Still further back in time, about 600–800 million years ago, the earth was probably covered entirely by ice, as mentioned earlier. A climate model should be able to account for all three types of climate—ice-covered globe, ice-free globe, and partially ice-covered globe—and explain transitions among such drastically different climates under solar inputs that have not fluctuated by more than 6% in hundreds of millions of years of earth's history.

Human influence on our climate (called anthropogenic climate forcing) is beginning to be noticeable two centuries after the Industrial Revolution. Of particular current concern is our increasing emission of carbon dioxide from the burning of fossil fuels. Most current climate models predict that this will lead to *global warming* through the greenhouse effect. How much the earth is predicted to warm is controversial, because it is currently still model-dependent and is under intense scientific and political debate. As of 2006, the United States has not joined the other nations in the Kyoto Protocol to curb future emissions of carbon dioxide because the administration feels that there is still scientific uncertainty. We will discuss these issues later in this chapter using a simple climate model.

8.2 Simple Climate Models

The simplest type of climate model is the energy balance models pioneered in 1969 separately by M. I. Budyko of the State Hydrological Institute in Leningrad and W. D. Sellers of the University of Arizona at Tucson, Arizona. These models try to predict the latitudinal distribution of surface temperature T, based on the concept that the energy the earth receives from the sun's radiation must balance the radiation the earth is losing to space by reemission at its temperature T. They also take into account the reflection of solar radiation back to space, the so-called *albedo* effect, by the ice and snow and by the clouds. For an ice-free planet, these models tend to give an annually averaged temperature at the equator of 46°C, and of −43°C at the pole. This is much warmer at the equator and much colder at the pole than our current values of 27°C and −13°C at the two locations, respectively. The subfreezing temperature at high latitudes is inconsistent with the prior assumption of an ice-free planet. Water must freeze at polar latitudes. Allowing for the formation of ice makes the problem much more interesting. Since ice reflects sunlight back to space more than land or ocean surfaces do, the earth is actually losing more heat—with less absorption of the sun's radiation—than if there were no ice cover. So it gets colder, consequently more ice forms, and the ice sheet advances equatorward. On the other hand, if, for some reason, the solar radiation is increased, ice melts a little near the ice edge, exposing more land, which absorbs more solar radiation, which makes the earth warmer, and more ice melts. The ice sheet retreats poleward. The albedo effect may serve to amplify any small changes to solar radiation that the earth receives in its orbit, depending on the effectiveness of dynamical transports. And since such orbital changes are known to be really small, the inherent instability of the ice–albedo feedback mechanism is therefore of much interest to climate scientists.

There are some minor differences between Budyko's and Sellers's models. We will discuss the Budyko model, as it has a simpler form of transport, which we can analyze using mathematics already introduced in a previous chapter.

Incoming Solar Radiation

The incoming solar radiation at the top of the earth's atmosphere is written as $Qs(y)$, where $y = \sin \theta$, with θ being the latitude. The latitudinal distribution function $s(y)$ is normalized so that its integral with respect to y from 0 to 1 is unity. Q is then the overall (integrated) total solar input into the atmosphere–ocean system. Its magnitude is 343 watts per square meter at present.

(*More geometrical details if you are interested*: Consider a very large sphere of radius r enclosing the sun at its center. The sun's radiation on the (inside) surface of the sphere, when integrated over the entire sphere, should be independent of r by conservation of energy. Let $S(r)$ be the radiation impinging on a unit area on that sphere. The area of a spherical surface at radius r is $4\pi r^2$. Since $S(r)4\pi r^2$ is independent of r, $S(r)$ therefore decreases with r as r^{-2}. The farther a planet is from the sun, the less radiation it receives per unit area. The value of $S(r)$ for r at the earth–sun distance is called the solar constant. The solar constant, along with detailed information on the solar radiation at various wavelengths of light and energy, has been measured by satellite since 1979. So we know the solar constant at the top of our atmosphere. At the mean earth–sun distance it is about 1,372 watts per square meter at present. Various parts of the earth receive more or less of the solar energy. On an annual average, the equator is closer to the sun than the poles, and so it receives more. The earth's rotational axis tilts (about 23.5° at present) from the normal to the elliptical plane of the earth's orbit. In January, the Southern Hemisphere is closer to the sun than the Northern Hemisphere, and vice versa in July. So there is a seasonal as well as a latitudinal variation of the incoming radiation. We shall consider here annual averages and deal with latitudinal variation only. The analytical formula for this, obtained from astronomical and geometrical calculations, is known but is complicated to write down. Nevertheless it has been tabulated; see Chylek and Coakley (1975). The rate of solar energy input per unit earth area is usually written in the form $Qs(y)$, where $y = \sin\theta$, with θ being the latitude. The total solar energy input is obtained by integrating over the surface of the earth of radius a:

$$\int_{-\pi/2}^{\pi/2} Qs(\sin\theta)2\pi a\cos\theta a d\theta = 4\pi a^2 Q \int_0^1 s(y)dy = 4\pi a^2 Q,$$

if the function $s(y)$ is normalized so that its integral from 0 to 1 is unity. The above integrated solar input should be equal to the solar flux intercepted by an area of the circular disk of the earth seen by the sun: $S\pi a^2$. Therefore $Q = S/4 = 343$ watts per square meter. The function $s(y)$ is uniformly approximated to within 2% by North (1975) to be

$$s(y) \cong 1 - S_2 P_2(y), \quad \text{where } S_2 = 0.482 \text{ and } P_2(y) = (3y^2 - 1)/2,$$

for the present obliquity of the earth's orbit. The obliquity is the angle between the earth's axis of rotation and the normal to the plane of its orbit around the sun. We shall consider $s(y)$ as known in our model to follow.)

Albedo

A fraction of the solar radiation is reflected back to space without being absorbed by the earth. Let $\alpha(y)$ denote the fraction reflected; α is called the *albedo* (from the Latin word *albus*, for whiteness; the word *albino* comes from the same root). The amount absorbed by the earth per unit area is therefore

$$Qs(y)(1 - \alpha(y)). \tag{8.1}$$

Outward Radiation

In the energy balance models, this absorbed solar energy is balanced at each latitude by reemission from the planet to space and the transport of heat by the atmosphere–ocean system from this latitude to another. Let $I(y)$ be the rate of energy emission by the earth per unit area. It is temperature dependent; the warmer the planet, the higher its rate of energy emission. It is given by

$$I = A + BT, \tag{8.2}$$

where T is the surface temperature in $^\circ$C. The constants A and B are chosen empirically based on the present climate. They are $A = 202$ watts per square meter, and $B = 1.90$ watts per square meter per $^\circ$C.

(*Some details of physics*: The earth's reemission of absorbed solar radiation is different from the reflection of solar radiation. The reflection of solar radiation occurs at the wavelength of the incident radiation, which contains mostly the ultraviolet and visible parts of the spectrum, without any transformation of the energy. In reemission, the earth–atmosphere–ocean system heats up after the absorption of the incoming solar radiation. From space the planet appears as a warm sphere that is radiating its energy to space at a rate characteristic of its emitting atmospheric layer, which is related to its surface temperature. For the temperatures we are considering, the reradiation occurs mostly at infrared wavelengths. A well-known law, the Stefan–Boltzmann law, states that for a black body (without an atmosphere), the rate of energy emission per unit surface area, $I(y)$, is proportional to the fourth power of the surface temperature T. It is written in the form $I(y) = \sigma T^4$, with T in absolute temperature and $\alpha = 5.6686 \times 10^{-8}$ watts per square meter per $^\circ K^4$ being the Stefan–Boltzmann constant. The earth is not a black body. In particular its atmosphere has several greenhouse gases, such as water vapor and carbon dioxide, that trap the infrared emission from the surface. This effect is taken into account in this simple model by multiplying σ by an emissivity fraction $\delta < 1$. A further difficulty is the nonlinear dependence of T, and this is dealt with in these simple models by linearizing (tangent approximation) about 0°C, which is $T_0 = 273^\circ$K. Thus,

$$I(y) = \delta\sigma T^4 \cong \delta\sigma T_0^4[1 + 4(T - T_0)/T_0] = A + B(T - T_0).$$

The second step above is a linear tangent approximation to the function T^4. This approximation then leads to (8.2). However, according to this tangent approximation, the constants should be $A = \delta\sigma T_0^4$ and $B/A = 4/T_0$. Different values of A and B have been used by various authors and they don't necessarily satisfy this relationship between A and B, because the δ may depend on T. Since $I(y)$ can now be measured directly by satellites as outgoing-longwave-radiation (OLR), one can directly fit a straight line to the measured data and obtain the parameters A and B. There is a very good correlation between OLR and the surface temperature, and it can be fitted to a straight line.

This procedure gives $A = 202$ watts per square meter and $B = 1.90$ watts per square meter per °C. (See Graves, Lee, and North [1993].)

Ice Dynamics

Ice forms from pure water when the temperature is below 0°C. However, permanent glaciers cannot be sustained until the annually averaged temperature is much colder, especially over the oceans. (If the annually averaged temperature is 0°C, it means that during summer the temperature is above freezing and the ice melts.) In the models of Budyko and Sellers the prescription is for an ice sheet to form when $T < T_c = -10$°C.

Let y_s be the location of the ice line, so that poleward of this latitude the earth is covered with ice and equatorward of this location it is ice-free. Since the albedo is higher in the ice-covered part of the earth, Budyko took the following form for $\alpha(y)$:

$$\alpha(y) = \begin{cases} \alpha_2 = 0.62 & y > y_s, \\ \alpha_1 = 0.32 & y < y_s. \end{cases} \tag{8.3}$$

At the ice boundary the temperature is taken to be T_c, i.e.,

$$T(y_s) = T_c.$$

Following Lindzen (1990), we assume the albedo at the ice edge to be the average of that on the ice side and on the ice-free side:

$$\alpha(y_s) = \alpha_0 = (\alpha_1 + \alpha_2)/2 = 0.47.$$

Transport

When a hot fluid is placed next to a cold fluid, heat is often exchanged in such a way as to make the temperature difference less. In ordinary fluids, such as water or air, this heat exchange is accomplished through either conduction or convection. Convection, involving the overturning of the fluid, which can carry heat directly from the hot spot to the cold spot, is often the more effective of the two mechanisms. Heat is transported by the earth's atmosphere–ocean system in a number of ways. In the tropical atmosphere, rapid vertical convection and the presence of a north–south overturning circulation (called the Hadley circulation) tend to smooth out the north–south temperature gradient. In the extratropical atmosphere, large-scale waves, in the form of cyclones, anticyclones, and storms, also act to transport heat from where it is warm to where it is colder. A detailed description of these processes will require a complex dynamical model involving many scales of motion.

In the simple model of Budyko, the transport processes are lumped into a simple relaxation term for the rate of change of heat energy due to all dynamical transport processes:

$$D(y) = C(\overline{T} - T), \qquad (8.4)$$

where $\overline{T}$ is the globally averaged temperature. The simple form in (8.4) satisfies the constraint that transport only moves heat from hot to cold while having no effect on the globally integrated temperature. If the local temperature at a particular latitude is greater than the global mean, heat will be taken out of that latitude. Conversely, if the local temperature is colder than the global mean, that latitude will gain heat. The empirical parameter C was assumed by Budyko to be $2.4B$ so that the solution can fit the present climate when the radiative parameters are taken to be the current climate values. Held and Suarez (1974) discussed how C and B can be evaluated from radiation and temperature measurements and suggested a value of $C = 2.1B$. Using the more updated value of the solar constant measured after 1979 using satellites, we choose $C = 1.6B$ (see later calculation of the present date ice-line location).

The Model Equation

We now construct a model equation. Our equation should say that the rate of change of earth's surface temperature should be equal to that due to the net absorption of solar energy input minus that due to earth's outward radiation, plus the heat gained or lost from transport. Thus:

$$\boxed{R\frac{\partial}{\partial t}T = F(T),} \qquad (8.5)$$

where $F(T) = Qs(y)(1 - \alpha(y)) - I(y) + D(y)$. The dependence of F on y and t is not displayed for convenience.

(Although we use the partial derivative with respect to t in Eq. (8.5) because T depends on both y and t, you can treat it the same as an ordinary derivative for all practical purposes. This is because there is no y-derivative in that equation; we can therefore treat y as another parameter, instead of as a second independent variable.) The parameter R on the left-hand side of (8.5) is the heat capacity of the earth, which is mostly determined by the heat capacity of the atmosphere and oceans. It is needed so that RT will have the dimension of energy, since the right-hand side contains energies. We will not need to specify a value for R. This time-dependent version of the Budyko equation was first used by Held and Suarez (1974), and further analyzed by many later authors, including Frederiksen (1976).

On an annual mean basis, the problem is symmetric about the equator, and so we will only need to consider the case of $y \geq 0$ after we assume the symmetry condition across the equator: $dT/dy = 0$ at $y = 0$. Under this symmetry condition, the global mean temperature is the same as the hemispherically averaged temperature, i.e.,

$$\overline{T} = \int_0^1 T(y)dy.$$

An equation governing the evolution of the global mean temperature can be obtained by integrating Eq. (8.5) hemispherically. It is

$$R\frac{d}{dt}\overline{T} = Q(1 - \overline{\alpha}) - A - B\overline{T}, \tag{8.6}$$

where $\overline{\alpha} = \int_0^1 s(y)\alpha(y)dy = \alpha_1 \int_0^{y_s} s(y)dy + \alpha_2 \int_{y_s}^1 s(y)dy.$ $\overline{\alpha} = \alpha_1$ for an ice-free globe; $\overline{\alpha} = \alpha_2$ for an ice-covered globe. For an earth partially covered by ice with the ice line at y_s, it is

$$\overline{\alpha} = \alpha_2 + (\alpha_1 - \alpha_2)y_s[1 - 0.241(y_s^2 - 1)].$$

For the present ice line, located at $y_s = 0.95$ (corresponding to $72°$N), $\overline{\alpha} = 0.33$, close to the ice-free albedo of 0.32.

8.3 The Equilibrium Solutions

We shall first seek the equilibrium solution T^* of Eq. (8.5) by setting its right-hand side to zero:

$$F(T^*) = Qs(y)(1 - \alpha(y)) - (A + BT^*) + C(\overline{T}^* - T^*) = 0. \tag{8.7}$$

This time-independent equation was first studied by Budyko. There are multiple equilibrium solutions depending on the extent of ice cover on the globe. The global mean temperature at equilibrium can be obtained directly by setting the right-hand side of Eq. (8.6) to zero:

$$\overline{T}^* = [Q(1 - \overline{\alpha}) - A]/B. \tag{8.8}$$

Substituting (8.8) into (8.7), we obtain the equilibrium solution:

$$T(y)^* = [C\overline{T}^* + Qs(y)(1 - \alpha(y)) - A]/(B + C)$$

$$= \frac{Q}{B + C}\left[s(y)(1 - \alpha(y)) + \frac{C}{B}(1 - \overline{\alpha})\right] - \frac{A}{B}. \qquad (8.9)$$

The location of the ice line is determined by evaluating (8.9) at y_s, where $T = T_c$:

$$T_c = \frac{Q}{B + C}\left[s(y_s)(1 - \alpha(y_s)) + \frac{C}{B}(1 - \overline{\alpha})\right] - \frac{A}{B}. \qquad (8.10)$$

(This equation is valid for $0 < y_s < 1$. When the ice line is at the equator, it cannot move any more equatorward even for higher Q. Similarly for the ice line at the pole; it cannot move any more poleward for smaller values of Q.)

   This equation yields the location of the ice line as a function of Q. Instead of solving (8.10), a cubic equation in y_s as a function of Q, one can alternatively solve for Q as a function of y_s, which is much easier. The result is shown in Figure 8.2.

Ice-Free Globe

   We first investigate the possibility of an ice-free solution. In that case, $\alpha(y) = \alpha_1 = 0.32$ everywhere. The solution in (8.9) becomes

$$T(y)^* = \frac{Q(1 - \alpha_1)}{B + C}\left[s(y) + \frac{C}{B}\right] - \frac{A}{B}. \qquad (8.11)$$

In order for it to be a self-consistent solution for an ice-free globe, solution (8.11) must be everywhere greater than T_c, including at the pole, the location of the minimum temperature. This condition is obtained by setting $T(1)^* > T_c$, thus yielding a restriction on the magnitudes of Q as a function of A, B, and C as

$$Q > \frac{(B + C)(T_c + A/B)}{(1 - \alpha_1)(s(1) + C/B)}.$$

For the parameter values given previously for the present climate, i.e., $A = 202$ watts per square meter, $B = 1.90$ watts per square meter per $°C$, and $C = 1.6B$, the condition that the polar temperature $T(1)^*$ in this ice-free scenario must be greater than T_c yields the condition on the

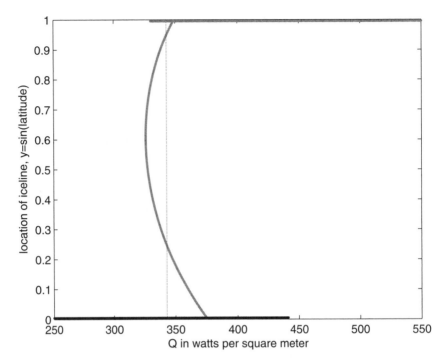

Figure 8.2. The location of the ice line, y_s, as a function of Q, obtained by evaluating for Q in Eq. (8.10) for various y_s. The vertical dotted line indicates the present climate at $Q = 343$. At this value of Q there are four possible locations of the ice line; the present location at $y_s = 0.95$ is one of the four possibilities. The top horizontal line is for the ice-free solution, while the lower horizontal line is for the snowball earth solution.

mean solar energy input of Q:

$$Q > 330 \text{ watts per square meter.}$$

Since our earth currently receives $Q = 343$ watts per square meter, this scenario of an ice-free globe is a distinct alternative climate under the present conditions provided that we can show that this equilibrium solution is stable. In such a climate, the globally averaged temperature is a warm 16°C:

$$\overline{T}^* = [Q(1 - \alpha_1) - A]/B = [343(0.68) - 202]/1.9 = 16°C.$$

Ice-Covered Globe

Similar to the previous section, we can investigate the possible solution for a completely ice-covered earth by setting the albedo to $\alpha(y) = \alpha_2 = 0.62$ everywhere.

The equilibrium temperature solution is, from Eq. (8.9):

$$T(y)^* = \frac{Q(1 - \alpha_2)}{B + C}\left[s(y) + \frac{C}{B}\right] - \frac{A}{B}. \tag{8.12}$$

Again, to be consistent with the prior assumption of an ice-covered globe, the temperature must everywhere be less than T_c, including at the equator, the location of maximum temperature. Using the same parameters as in our current climate, we find that a completely glaciated globe is a possibility if the solar input drops below a threshold value given by

$$Q < \frac{(B + C)(T_c + A/B)}{(1 - \alpha_2)(s(0) + C/B)},$$

$Q < 441$ watts per square meter.

Since we currently receive even less than this threshold value—our current Q is 343 watts per square meter—our earth might alternatively be *totally ice covered* if this equilibrium turns out to be stable. In such a climate, the globally averaged temperature is a frigid *minus* 38°C:

$$\overline{T}^* = [Q(1 - \alpha_2) - A]/B = [343(0.38) - 202]/1.9 = -38°C.$$

Partially Ice-Covered Globe

The more general solution is a globe partially covered by ice. The mathematics is slightly more involved, but still straightforward. To find the global mean temperature we can either use Eq. (8.8) or evaluate Eq. (8.7) at the ice edge. The latter procedure yields

$$\overline{T}^* = A/C + (1 + B/C)T_c - Qs(y_s)(1 - \alpha_0)/C, \tag{8.13}$$

where $\alpha_0 = \alpha(y_s)$.

Solving Eq. (8.7) separately for the ice-covered part and the ice-free part of the globe, we find

$$T(y)^* = T_1(y) = [Q(1 - \alpha_1)s(y) + C\overline{T}^* - A]/(B + C) \quad \text{for } y < y_s,$$

$$T(y)^* = T_2(y) = [Q(1 - \alpha_2)s(y) + C\overline{T}^* - A]/(B + C) \quad \text{for } y > y_s.$$

We substitute (8.13) into these expressions and find that we can write the above solution in the following compact form (due to Frederiksen [1976]):

$$T_i(y) = T_c + \frac{Q}{B + C}[s(y)(1 - \alpha_i) - s(y_s)(1 - \alpha_0)], \quad i = 0, 1, 2. \tag{8.14}$$

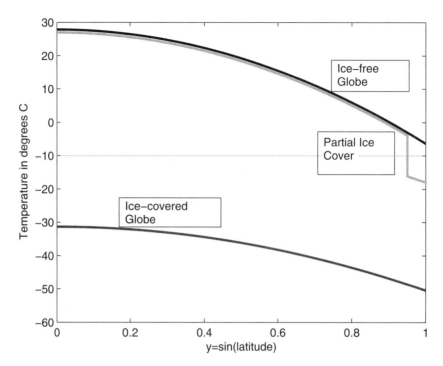

Figure 8.3. Equilibrium temperature as a function of y for current climate parameters ($Q = 343$, $A = 202$, $B = 1.90$, and $C = 1.6B$). Note that for the same solar constant as in the current climate, it is possible to have an ice-free globe (top curve), an ice-covered globe (bottom curve), and the current climate, which is a partially ice-covered globe with the ice line at $y = 0.95$ (the intermediate, discontinuous curve).

For $Q = 343$ watts per square meter and for the ice line located at $72°$ latitude, (8.14) gives the temperature distribution for our "current" climate in this simple model. This is plotted in Figure 8.3. The globally averaged temperature of this equilibrium solution is, from either (8.8) or (8.13),

$$\overline{T}^* = [343(1 - 0.33) - 202]/1.9 = 15°C,$$

which is quite close to the observed global mean temperature currently.

Multiple Equilibria

We see from the above results that there exist multiple equilibria under the same set of parameter values. For example, for the same current solar forcing, we can have either the current climate with a global mean temperature of $15°C$, a completely ice-covered earth with a global mean temperature of $-38°C$, or a completely ice-free earth with a

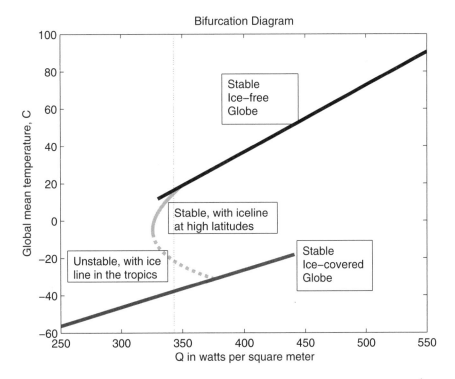

Figure 8.4. Diagram of global mean equilibrium temperature vs. Q, which is $\frac{1}{4}$ of the solar constant. The current value $Q = 343$ is indicated with a vertical dotted line. At this value of Q there are four equilibrium solutions. The top one is ice free, the lower one is ice covered, and the middle two have partial ice cover. The current climate has the ice line at high latitudes, and there is another equilibrium with the ice line in the tropics (which turns out to be unstable). The unstable equilibrium is denoted by a dashed line.

global mean temperature of 16°C. These multiple branches of solutions are plotted in Figure 8.4. Some branches are stable while others are not. The stability of these equilibria is discussed next.

8.4 Stability

Let Eq. (8.6) be written as

$$R\frac{d}{dt}\overline{T} = G(\overline{T}),$$

where we have symbolically let $G(\overline{T}) = Q(1 - \overline{\alpha}) - A - B\overline{T}$. The equilibrium solutions are given by the zeros of G and have been denoted with an asterisk. We perturb the temperature slightly from that equilibrium and write

$$\overline{T} = \overline{T}^* + u(t).$$

We furthermore expand G in a Taylor series and drop terms of order u^2 and higher, since if u is small, u^2 is even smaller (this process of approximating a nonlinear function by a linear function is called *linearization*):

$$G(\overline{T}) = G(\overline{T}^* + u) \approx G(\overline{T}^*) + \frac{dG}{d\overline{T}}(\overline{T}^*)u = \frac{dG}{d\overline{T}}(\overline{T}^*)u,$$

where

$$\frac{dG}{d\overline{T}}(\overline{T}^*) = -B - Q\frac{d\overline{\alpha}}{d\overline{T}}(\overline{T}^*)$$

from the expression for G. Furthermore, differentiating the equilibrium solution (8.8) with respect to itself, we get

$$B = \frac{dQ}{d\overline{T}^*}(1 - \overline{\alpha}) - Q\frac{d\overline{\alpha}}{d\overline{T}^*}.$$

Therefore,

$$\frac{dG}{d\overline{T}}(\overline{T}^*) = -(1 - \overline{\alpha})\frac{dQ}{d\overline{T}^*}.$$

The Slope-Stability Theorem

Thus the time-dependent equation governing the temperature perturbation is

$$R\frac{d}{dt}u(t) = -\gamma u(t),$$

where we have let $\gamma \equiv (1 - \overline{\alpha})\frac{dQ}{d\overline{T}}$. Its sign depends on the sign of $\frac{dQ}{d\overline{T}}$.
The solution to the equation above is

$$u(t) = u(0)\exp\left\{-\frac{\gamma}{R}t\right\}. \tag{8.15}$$

The perturbation will decay in time if γ is positive. In this case the equilibrium is stable to small perturbations. If γ is negative, the small perturbation will grow larger; the equilibrium is unstable to small perturbations. We now have obtained the so-called slope-stability theorem

(Cahalan and North (1979)):

$$\frac{dQ}{d\overline{T}^*} > 0 : \text{ stable,}$$

$$\frac{dQ}{d\overline{T}^*} < 0 : \text{ unstable.}$$

This result was first obtained by Budyko (1972) using intuitive arguments.

One can examine the equilibrium diagram in Figure 8.4 and see that the equilibrium solution branch with the positive slope is stable, while that with the negative slope is unstable. With the slope of the equilibrium diagram yielding information on the stability of the equilibrium solution itself, this diagram is thus seen to be doubly useful. We see that the branch for the ice-free solution and the branch for the ice-covered solution have positive slopes, and therefore we conclude that these two scenarios are stable. For a globe partially covered by ice, it appears that once the ice sheet covers about half the earth's area it becomes unstable. Our current climate with the ice sheet at high latitudes is stable.

Alternatively, one can differentiate the equilibrium temperature with respect to Q and obtain the slope analytically. This will be done in the following two subsections. They can be skipped if you are satisfied with the numerical/graphical solution depicted in Figure 8.4.

The Stability of the Ice-Free and Ice-Covered Globes

Examining the equilibrium diagram in Figure 8.4, we see that both the ice-free globe and the ice-covered globe correspond to stable equilibria. One can also show explicitly, by differentiating (8.8) with respect to Q, that

$$\frac{d\overline{T}^*}{dQ} = \frac{(1 - \alpha_i)}{B} > 0$$

for either the ice-free case ($i = 1$) or the ice-covered case ($i = 2$), thus satisfying the condition for stability. Note that this stability condition is independent of many factors affecting the current climate and independent of C, hence independent of our parameterization of dynamical transport, the weakest part of the model.

Stability and Instability of the Partially Ice-Covered Globe

For the case of a partially ice-covered globe, we again differentiate Eq. (8.8) with respect to Q, but this time we note that $\overline{\alpha}$ is a function of y_s, which depends on Q:

$$B\frac{d\overline{T}^*}{dQ} = (1 - \overline{\alpha}) + Q\left(-\frac{d\overline{\alpha}}{dy_s}\right)\frac{dy_s}{dQ}.$$ (8.16)

We know

$$\frac{d\overline{\alpha}}{dy_s} = -(\alpha_2 - \alpha_1)[1 - 0.482y_s - 0.241(y_s^2 - 1)]$$

is always negative. This is consistent with our intuition that as the ice sheet retreats poleward, exposing darker surfaces, the overall albedo of the earth will decrease. It then follows that if dy_s/dQ is positive (i.e., the ice line would retreat poleward with an increase of solar constant), (8.16) will be positive and the equilibrium solution will be stable. Differentiating (8.10) with respect to y_s, we find

$$0 = \frac{dQ}{dy_s}\left[s(y_s)(1 - \alpha_0) + \frac{C}{B}(1 - \overline{\alpha})\right] + Q\left[-3 \cdot 0.482y_s(1 - \alpha_0) - \frac{C}{B}\frac{d\overline{\alpha}}{dy_s}\right].$$

Rearranging,

$$\frac{1}{Q}\frac{dQ}{dy_s} = \left[1.45y_s(1 - \alpha_0) + \frac{C}{B}\frac{d\overline{\alpha}}{dy_s}\right] \bigg/ \left[s(y_s)(1 - \alpha_0) + \frac{C}{B}(1 - \overline{\alpha})\right].$$ (8.17)

Substituting (8.17) into (8.16), we find

$$B\frac{d\overline{T}^*}{dQ} = (1 - \overline{\alpha}) + \left(-\frac{d\overline{\alpha}}{dy_s}\right)\frac{[s(y_s)(1 - \alpha_0) + \frac{C}{B}(1 - \overline{\alpha})]}{[1.45y_s(1 - \alpha_0) + \frac{C}{B}\frac{d\overline{\alpha}}{dy_s}]}.$$

Therefore the decay rate in (8.15) is

$$\gamma \equiv (1 - \overline{\alpha})\frac{dQ}{d\overline{T}^*} = \frac{[1.45y_s(1 - \alpha_0) + \frac{C}{B}\frac{d\overline{\alpha}}{dy_s}]/B}{\left(-\frac{d\overline{\alpha}}{dy_s}\right)s(y_s)(1 - \alpha_0) + 1.45y_s(1 - \alpha_0)(1 - \overline{\alpha})}.$$ (8.18)

So γ changes sign when the numerator in the above expression changes sign. This occurs when

$$1.45(1 - \alpha_0)y_s = \frac{C}{B}\left(-\frac{d\overline{\alpha}}{dy_s}\right).$$ (8.19)

(Note that the radiative equilibrium solution, obtained by setting the dynamical transport C to zero, yields a positive numerator. Hence (8.18) is always positive for that solution. The radiative equilibrium solution is stable wherever the ice line is positioned. It is the dynamical transport that destabilizes the ice-albedo feedback.) In the presence of nonzero transport, C, there are two roots to the quadratic equation (8.19), one positive and one negative. The positive root is

$$y_s = -\left[1 + 3\frac{(1-\alpha_0)B}{(\alpha_2-\alpha_1)C}\right] + \sqrt{\left[1 + 3\frac{(1-\alpha_0)B}{(\alpha_2-\alpha_1)C}\right]^2 + 5.15} \approx 0.56;$$

(8.20)

that is, about $34°$ latitude. Equation (8.18) is positive if the ice line is located poleward of this latitude, and the equilibrium solution is stable.

Luckily our present climate, with the ice line located at $72°$ latitude, is stable, according to this simple model. One way to gauge how complacent we can be is to ask: By how many percentage points can Q change from the value of our current climate before our climate is moved from the stable equilibrium to the unstable equilibrium? In other words, how much must Q change to move the ice line from $72°$ to $34°$ of latitude? This problem is left to exercise 5. You will be surprised by how small this value is. Once at $34°$, the ice-albedo feedback will initiate a runaway freeze.

Since the stability property depends critically on dynamical transport, and our treatment of transport is admittedly very crude, the above result may change with better models. Nevertheless, about the same conclusions were obtained by North (1975) using a model with diffusive heat transport (see exercise 6), including the result that the ice-free globe, the ice-covered globe, and the present climate are stable, and that the globe becomes unstable when the ice sheet advances to near the tropics. (Some more recent general circulation models incorporating detailed atmospheric circulations and ice dynamics appear to show that a very narrow band of water on the equator may remain ice free even when our simple model predicts a snowball earth. This open water might have provided a refuge for multicellular animals through the deep freeze. See Hyde et al. [2000].)

How Does a Snowball Earth End?

If for some reason the ice sheet advances past $34°$, the solution will become unstable. The ice sheet will then advance all the way to the equator, reaching the stable equilibrium of a snowball earth. Considering the fact that the sun's output 600 million years ago was 6% less than the present value, we see that the possibility of having one really cold winter (for one reason or another) with the ice sheet advancing into the tropics is rather real.

Once the earth is completely glaciated, the above simple analysis suggests that it would remain so. The global temperature would plummet to less than $-42°C$. The earth could not escape its ice-encased tomb unless the solar constant were increased by more than 40% (from $Q = 322$ to $Q > 450$ watts per square meter), which we know did not happen.

On the other hand, for the same solar input, the atmosphere could have warmed up by increasing its greenhouse effect, which lowers its emissivity δ. This has the effect of lowering the parameters A and B, which are here calibrated using the present value of emissivity. Caldeira and Kastings (1992) investigated the effect of varying amounts of carbon dioxide concentration in the atmosphere, measured by its partial pressure, pCO_2, on the OLR: $I = A + BT$. Using results from 2,000 runs of radiative equilibrium calculations with different carbon dioxide partial pressures, they fitted the constants A and B as a function of $\varphi = \ln(pCO_2/(pCO_2)_{ref})$, where $(pCO_2)_{ref}$ is a reference value corresponding to the present value of CO_2 at 300 parts per million:

$$A = -326.4 + 9.161\varphi - 3.164\varphi^2 + 0.5468\varphi^3 \text{ watts meter}^{-2},$$

$$B = 1.953 - 0.04866\varphi + 0.01309\varphi^2 - 0.002577\varphi^3 \text{ watts meter}^{-2\circ}K^{-1}.$$

$$(8.21)$$

Setting $\varphi = 0$ should give close to our current value of A and B. (Note that the authors used degrees K for their T instead of our degrees C, and so one should add $273B$ to their A to convert into our A.)

Let h be the factor by which A and B must be reduced from their current values so that

$$Q/h > 441 \text{ watts per square meter.}$$

Therefore h must be less than 73% if Q is at 322 watts per square meter. It was estimated that the needed carbon dioxide concentration in the atmosphere would have been 400 times the present concentration in order to initiate a meltdown!

8.5 Evidence of a Snowball Earth and Its Fiery End

Brian Harland of Cambridge University was the first to suggest, in the early 1960s, that the earth experienced a "great infra-Cambrian glaciation" 600 million years ago. He came to this conclusion by noting that glacial deposits were found in rocks dated to that period (called the Neoproterozoic period by geologists) across virtually every continent on earth. In particular, Harland found glacial deposits within types of marine sedimentary strata characteristic of low latitudes. There has not been evidence of ice at sea level at the equator again since that time.

Figure 8.5. Daniel Schrag (left) and Paul Hoffman (right) point to a layer of abrupt cap carbonate rocks above a layer of glacial marine dropstones in Namibia. (Photo by Gabrielle Walker, courtesy of Paul Hoffman.)

Today we find glaciers at the equator only more than 5,000 meters above sea level, above the Andes and Mt. Kilimanjaro. They came down to no lower than 4,000 meters above sea level during the last ice age.

In 1992, Joseph Kirschvink of Caltech suggested that carbon dioxide supplied by volcanoes might have been what was needed for the earth to escape from its icy tomb, which otherwise might have been permanent, as we have inferred from the Budyko model. On an ice-covered earth, the normal process of removing carbon dioxide from the air would have been absent, while the input from volcanoes continued. There would not have been any evaporation and thus no rain or even snow in such a cold climate. Rain acts in our present climate to wash carbon dioxide from the air, and the weathering of silicate rocks on land converts carbon dioxide to bicarbonate, which, when washed to the oceans, becomes carbonate sediments. It was estimated by Hoffman et al. (1998) that with such a removal process shut down, it would have taken the volcanoes ten million years to build up the carbon dioxide level in the air—400 times the present level—needed to initiate a hyper-greenhouse that was capable of deglaciating the snowball earth.

From their field observations of rock cliffs in Namibia (see Figure 8.5) and elsewhere, Paul Hoffman and his colleagues found that Neoproterozoic glacial deposits are almost always capped by carbonate rocks, which typically form in warm water, and the transition from glacial deposits to the cap carbonates is abrupt, occurring in perhaps a few thousand years. In their 1998 *Science* article, Hoffman et al. pieced this and other (isotopic) evidence together and suggested that the cap carbonate sediments must have formed in the aftermath of the snowball earth: rain in an atmosphere high in carbon dioxide would be in the form of acid rain, accelerating the erosion of rocks on exposed land. The sediments were then deposited at the bottom of the shallow seas, forming the observed cap carbonate rocks. Thus there is evidence of both a snowball earth and its abrupt end in a hot greenhouse.

The current debate is related to the question of whether the earth actually experienced a "hard snowball" (with the planet completely covered by ice) or a "soft snowball" (with a strip of open ocean near the equator). Pierrehumbert (2005) questioned whether it is at all possible to deglaciate a hard snowball earth by increasing carbon dioxide.

8.6 The Global Warming Controversy

We now turn to a problem closer to our time. We have already gained some sense from discussions on paleo-climate that our planet has a very sensitive climate system. Small radiative perturbations can lead to dramatic changes in our climate through feedback processes. We have discussed the ice-albedo feedback process in relation to the onset of a snowball earth and briefly mentioned the greenhouse effect of carbon dioxide from cumulative emissions by volcanoes in melting the ice. We now want to study the greenhouse effect more quantitatively (Figure 8.6).

Carbon dioxide is but one of many greenhouse gases naturally occurring in our atmosphere, the other greenhouse gases being methane, nitrous oxide, and, more importantly, water vapor. These greenhouse gases are what is responsible for our current global temperature of $15°C$. Without them, our global temperature would be a chilly $-17°C$. Prior to the Industrial Revolution, carbon dioxide concentration was probably around 280 parts per million of air, but it has since increased rapidly. In the United States, carbon dioxide constitutes about 80% of all anthropogenic emissions of greenhouse gases and is currently increasing at the rate of 2% per year.

There is no controversy concerning the fact that the carbon dioxide concentration in the atmosphere is increasing steadily. Measurements at the pristine mountaintop of Mauna Loa show in Figure 8.7 a steady increase from 310 parts per million in air in 1958 to our current concentration of 375 ppm. (There is a pronounced seasonal cycle in the carbon dioxide emissions, as plants suck up more carbon dioxide during the summer growing season. In fall the decay of leaves releases some carbon dioxide back to the atmosphere. We are interested in the

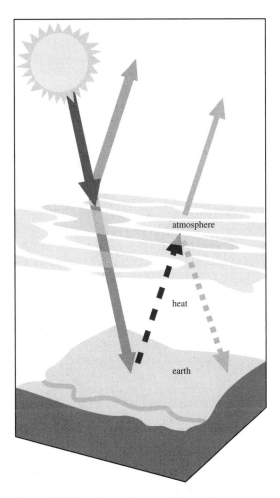

Figure 8.6. An atmosphere with greenhouse gases such as carbon dioxide traps more of the outgoing radiation from earth's reemission, increasing the warming.

annually averaged value, denoted by the black line through the seasonal fluctuations.) There is even evidence from ice cores (Figure 8.8) that the atmospheric carbon dioxide concentration hovered around 280 ppm for over a thousand years prior to the 1800s, and has increased rapidly since.

Like a greenhouse, which admits short-wave radiation from the sun through its glass but traps within the greenhouse the infrared reemission from inside the greenhouse, the greenhouse gases in the atmosphere warm the lower atmosphere of the earth by keeping in more of the infrared reemission from the ground (see Figure 8.6). It has

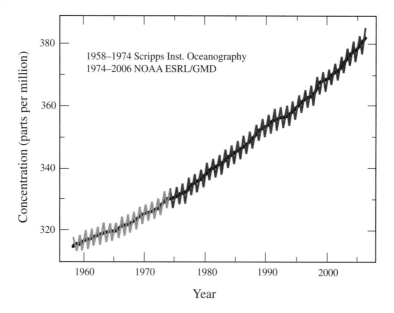

Figure 8.7. Measurement of atmospheric carbon dioxide at Mauna Loa in Hawaii. The vertical axis is its concentration in air in parts per million. The horizontal axis is the year.

been estimated that a doubling of atmospheric carbon dioxide is equivalent to an additional net radiative heating of the lower atmosphere of 2.6 watts per square meter (more specifically, $\delta Q(1 - \alpha) \sim 2.6$ watts per square meter) (Hansen et al., 2005).

The controversy centers around the following quantitative question: If the carbon dioxide concentration in the atmosphere is doubled, say, from its preindustrial value of 280 ppm, how much warmer will the global temperature be? This question can be phrased either as an equilibrium response or as a time-dependent response. A back-of-the-envelope estimate of the equilibrium response can be obtained from $\delta T \sim \delta Q(1 - \alpha)/B \sim 2.6/1.9 \sim 1.4°C$. Yet the model predictions of global warming due to a doubling of carbon dioxide span an uncomfortably large range: from 1.5° to 4.5°C. Despite intense efforts of hundreds of climate modelers and two Intergovernmental Panels on Climate Change (IPCC, 1990; IPCC, 2001), this large range of uncertainty remained almost unchanged for more than two decades. While a global warming of 4.5°C may be alarming and a cause for concern and calls for immediate action, an eventual warming of 1.5° may be, to some people, more benign to human society. Given the high cost of the proposed remedy (involving drastic curbs on the burning of fossil fuels) to nations' industrial production and development, the large scientific

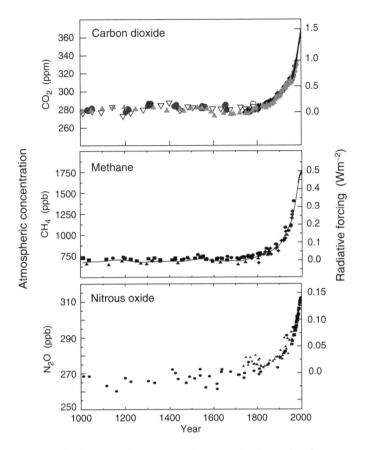

Figure 8.8. Records of atmospheric concentrations of carbon dioxide, methane, and nitrous oxide for the past 1,000 years. Ice core and firn data from several sites in Antarctica and Greenland (shown by different symbols) are supplemented by direct atmospheric samples over the past few decades. The estimated radiative forcing from these greenhouse gas changes is indicated on the right-hand scale. (From IPCC [2001]; courtesy of Intergovernmental Panel on Climate Change.)

uncertainty from model predictions fuels political debates on whether nations should undertake immediate action to curb carbon dioxide emission despite the cost.

Although there is still debate and uncertainty, it appears that the global temperature has warmed by $0.6° \pm 0.2°C$ since 1880 (IPCC, 2001). It has also been estimated that the increase in the greenhouse gases has produced an additional net radiative heating of $(1 - \alpha)\delta Q(t) \sim 1.80 \pm 0.85$ watts m^{-2} (IPCC, 2001; Hansen et al., 2005). The rather large uncertainty of this estimate is due to the uncertain effects of aerosol (soot) pollution, which, depending on particle size, generally tend to cool the surface by dimming the sunshine. The same back-of-the-envelope calculation would show that we should have warmed by $\delta T \sim \delta Q(1 - \alpha)/B \sim 1.8/1.9 \sim 0.95°C$, which is close, given the indicated uncertainties in the figures, to the observed warming that has already taken place. So if this estimate is consistent with the data on the global warming that has been observed so far, by extrapolation we should get $1.4°C$ of total warming when carbon dioxide is doubled. Why is there still this controversy on the predicted range of warming? The problem is actually more complicated. It turns out that there is a large difference between equilibrium warming—the warming that the climate settles down to eventually—and the time-dependent warming while the greenhouse gases are still increasing. There is also a large difference between the back-of-the-envelope estimates we have used above and the more comprehensive calculations involving a range of feedback mechanisms.

In the next section we will use the simple climate model developed so far to try to understand the source of the uncertainty involving the feedback processes and discuss a possible way to reduce it. There is even greater uncertainty concerning the time-dependent solution, because it involves the thermal inertia of the atmosphere and oceans. There is, however, a greater need to understand the time-dependent solution, because it is more relevant.

8.7 A Simple Equation for Climate Perturbation

We again use the time-dependent, annually averaged energy balance climate model of Held and Suarez (1974) governing the near-surface atmospheric temperature $T(y, t)$:

$$R\frac{\partial}{\partial t}T = Qs(y)(1 - \alpha(y)) - (A + BT) + \nabla \cdot (\text{heatflux}), \qquad (8.22)$$

where Q is $1/4$ of the solar constant, $s(y)$ its distribution with respect to latitude, globally normalized to unity, $y = \sin(\text{latitude})$, and $\alpha(y)$ is the albedo—the fraction of the sun's radiation reflected back to space by clouds and the earth's surface. $(A + BT)$ is the linearized form of the infrared emission of the earth to space fitted from observational data on outgoing longwave radiation (Graves, Lee, and North 1993), with $A = 202$ watts m^{-2} and $B = 1.90$ watts m^{-2} C^{-1} for our current climate. They are temperature dependent when our current climate is perturbed. The parameter R in Eq. (8.22) represents the thermal capacity of the atmosphere–ocean climate system. Its value is uncertain and that has prevented the use of the time-dependent version of this equation.

Dynamical transport of heat is written in the more general form of a divergence of heat fluxes. Nevertheless, its global average vanishes, as in Eq. (8.6). We will consider here the global average to Eq. (8.22), which is (the same as Eq. (8.6) but repeated here for convenience):

$$R\frac{\partial}{\partial t}\overline{T} = Q(1 - \overline{\alpha}) - (A + B\overline{T}), \qquad (8.23)$$

where an overbar denotes the global average and $\overline{\alpha} = \frac{1}{2}\int_{-1}^{1}\alpha(y)s(y)dy$ is the weighted global average albedo. The overbar is henceforth dropped for convenience.

Considering a small radiative perturbation δQ in $Q = Q_0 + \delta Q$, we find that the equation governing the small temperature perturbation can be obtained from the first variation of the above equation. We write

$$T = T_0 + \delta T,$$

where T_0 is the unperturbed temperature and δT is the perturbation temperature response to the perturbation in heating δQ. Linearizing Eq. (8.23) (using a Taylor series expansion in T about T_0) then leads to the following perturbation equation:

$$R\frac{\partial}{\partial t}\delta T = (1 - \alpha)\delta Q - B\delta T - \left(\frac{\partial}{\partial T}A\right)_0 \delta T - \left(T\frac{\partial}{\partial T}B\right)_0 \delta T - \left(Q\frac{\partial}{\partial T}\alpha\right)_0 \delta T.$$

This can be rewritten as

$$B\tau\frac{\partial}{\partial t}\delta T = (1 - \alpha)\delta Q - B\delta T/g,$$

where

$$\tau \equiv R/B,$$
$$g = 1/(1 - f), \qquad (8.24)$$
$$f = f_1 + f_2,$$
$$f_1 = \left(-T\frac{\partial}{\partial T}B - \frac{\partial}{\partial T}A\right)_0,$$
$$f_2 = \left(-Q\frac{\partial}{\partial T}\alpha\right)_0.$$

The parameter τ measures the time scale of the climate system's thermal inertia. It involves not only the thermal inertia of the atmosphere but also the much larger inertia of the oceans. Its value is uncertain. The factor g is the controversial climate gain; it amplifies any response to radiative perturbation by a factor g (see below). f_1 incorporates the effect of the water-vapor feedback and f_2 that of ice and snow albedo feedback. Cloud feedback has effects on both f_1 and f_2. The back-of-the-envelope calculation we used previously ignored the temperature dependence of the radiative parameters.

The water-vapor feedback factor is potentially the largest and therefore the most controversial. The cloud feedback is the most uncertain; even its sign is under debate. Note that various feedback processes can be superimposed in f (but not in g).

Water-Vapor Feedback

When the surface warms, it is natural to expect that there will be more evaporation and hence that more water vapor will be present in the atmosphere. Since water vapor is a natural greenhouse gas, one expects that the initial warming may be amplified, i.e., that the factor g should be greater than 1. In one of the earliest models of global warming, in 1967, Syukuro Manabe and Richard Wetherald of Princeton's Geophysical Fluid Dynamics Laboratory made the simplifying assumption that the relative humidity of the atmosphere remains unchanged when the atmosphere warms. This is in effect saying that the atmosphere can hold more water vapor if it is warmer. The presence of this additional greenhouse gas (i.e., water vapor) would amplify the initial warming and double it (see Hartmann, 1994). That is, the climate gain factor is $g_1 \sim 2$ due to the water-vapor feedback alone. This implies a feedback factor of $f_1 \sim 0.5$. This result appears to have stood the test of time. Most modern models yield water-vapor amounts consistent with this prediction. However, that most models tend to have similar water-vapor feedback factors does not necessarily mean that they are all correct.

Cloud Feedback

Cloud tops reflect visible sunlight back to space. Therefore, more clouds imply higher albedo and cooling. However, clouds also behave like greenhouse gases in trapping infrared radiation from below. Clouds are actually the second most important greenhouse gas, after water vapor but ahead of carbon dioxide. The cancellation of the albedo effect and the greenhouse effect of the clouds differs in different climate models, depending on the height and the type of clouds. As a consequence, even the sign of the cloud feedback is uncertain, although typical values in some climate models are around ~ 0.1 for the f factor. It is probably close to zero.

Ice and Snow Albedo Feedback

As the surface warms, snow or ice melts, exposing the darker surface underneath, thus lowering the albedo and increasing the absorption of the sun's radiation. This is a positive feedback process and is probably more important at high latitudes than at low latitudes. It may explain the higher sensitivity of the polar latitudes to global warming. On a globally averaged basis it is probably between 0.1 and 0.2 for the f factor.

Total Climate Gain
Adding all the feedback processes yields $f \sim 0.7$ in most climate models. This then yields a climate gain factor of $g = 1/(1 - f) \sim 3$. As noted previously, this number is uncertain.

Using Observation to Infer Climate Gain
The sun's radiation is observed to vary slightly over an 11-year cycle. This is related to the appearance of darker sunspots on the surface of the sun and the accompanying bright faculae. Sunspots have been observed since ancient times, but an accurate measurement of their radiative variation was not available until recently when, starting in 1979, satellites could measure the solar constant S above the earth's atmosphere. It was found that the solar constant varies by about 0.06% over a solar cycle. The atmosphere's response near the surface to this solar cycle variation has also been measured to be about 0.2°C on a global average. This information can be used to infer a climate gain factor. This is left as an exercise (in exercises 7 and 8). This leads to $g \sim 3$.

One can see the effect of g on the climate response even without solving Eq. (8.24). When multiplied by g throughout, that equation becomes

$$(g\tau)\frac{\partial}{\partial t}\delta T = (1 - \alpha)(\delta Qg)/B - \delta T.$$

This equation shows that the effective radiative forcing is $g\delta Q$, and the effective time scale involved in the climate's response is $g\tau$. Thus, the magnitude of the response to radiative forcing may be amplified by a factor of $g \sim 3$ because of the presence of feedback processes, but the time it takes to realize that larger response may be three times as long. This result is demonstrated below with explicit solutions.

8.8 Solutions

Equilibrium Global Warming
Setting the time derivative to zero, the steady state solution to (8.24) is

$$(\delta T)_{eq} = \frac{(1 - \alpha)\delta Q}{B}g. \tag{8.25}$$

The solution prominently shows the climate gain factor g in amplifying the equilibrium response to a given radiative forcing. For an "adjusted radiative forcing" due to doubling CO_2 of $(1 - \alpha)\delta Q = 2.6$ watts m^{-2}, the expected global warming is 1.4°C without the amplifying factor but 4.1°C with the amplifying factor of $g = 3$.

The range of current model predictions of 1.5°–4.5°C indicates that the various models have different feedback mechanisms and that their climate gain factor ranges from $g \sim 1$ to 3.

Time-Dependent Global Warming

Growth Phase

As a model for the increase in greenhouse gases, we assume that their radiative forcing has increased linearly since 1880, which we call $t = 0$:

$$(1 - \alpha)\delta Q(t) = bt, \quad \text{for } t > 0$$

and

$$\delta Q(t) = 0, \quad \text{for } t < 0. \tag{8.26}$$

This is the model considered by Hartmann (1994). It leads to an approximately linear increase in global warming. Given the recent accelerated warming, a model giving rise to an exponentially increasing temperature may be more appropriate. This latter model is discussed in exercises 9 and 10. Staying with Eq. (8.26), we now obtain the solution to the time-dependent equation (8.24). In view of the form of the forcing term, we assume the solution to consist of a homogeneous solution plus a particular solution (see Appendix A for a review). The particular solution is of the form $\delta T_{particular} = at - c$, and the homogeneous solution is of the form $\delta T_{homogeneous} = c \exp\{-t/(g\tau)\}$. (The two c's are of opposite sign so as to satisfy the initial condition that the total temperature perturbation be zero at $t = 0$.) The constants a and c are found by substituting this assumed solution into Eq. (8.24). This yields, for the sum of homogeneous plus particular solutions,

$$\delta T(t) = (bg/B)(t - g\tau) + (bg^2\tau/B) \exp\{-t/g\tau\}, \quad \text{for } t > 0. \tag{8.27}$$

The solution can be written in the following more interesting form:

$$\delta T(t) = \frac{(1 - \alpha)\delta Q(t - \Delta)}{B} g, \tag{8.28}$$

where

$$\Delta \equiv g\tau \left(1 - \exp\left\{-\frac{t}{g\tau}\right\}\right)$$

is the time delay. The delay is initially zero at $t = 0$ and increases steadily to a maximum of $g\tau$ for $t \gg g\tau$. The solution, Eq. (8.28), looks just like the equilibrium solution (8.25), except that it is evaluated at time t using the value of the radiative forcing at time $t - \Delta$. We call this the quasi-equilibrium solution with delay.

Curbs in Effect

Suppose that at some $t = t_s$ in the future, all nations decide to implement a curb on emissions of greenhouse gases. For simplicity, we assume that the emission curbs are such that the concentration of the greenhouse gases in the atmosphere remains constant:

$$\delta Q(t) = \text{constant} \quad \text{for } t > t_s. \tag{8.29}$$

The solution for the constant forcing case can be found using a particular solution, which is a constant. This constant is found by substituting this trial particular solution into Eq. (8.24): $\delta T_{\text{particular}} = (1 - \alpha)\delta Q(t_s)g/B$. The homogeneous solution is the same as before, $\delta T_{\text{homogeneous}} = c \exp\{-t/(g\tau)\}$, but now the constant c needs to be evaluated so that the solution at t_s matches that from Eq. (8.28). This yields

$$\delta T(t) = \frac{(1 - \alpha)\delta Q(t_s)g}{B} \left(1 - \left(\frac{g\tau}{t_s}\right) \exp\left\{-\frac{t - t_s}{g\tau}\right\}\right). \tag{8.30}$$

We see that eventually the warming will approach the equilibrium value predicted by Eq. (8.25) and that warming will be amplified by the climate gain factor g. However, it takes a time longer than $g\tau$ to reach that equilibrium. We now have a conclusion that is consistent with what other scientists have found using more complex computer model simulations (see Hansen et al., 1985) and is rather general:

The more sensitive the climate response (the larger the climate gain factor), the larger the global warming at equilibrium will be. However, it also takes longer to reach that equilibrium.

Next we will try to determine how long is "long."

Thermal Inertia of the Atmosphere–Ocean System

Before we can gain any insight from the time-dependent solution, we need to estimate the thermal capacity $R = B\tau$ of the atmosphere–ocean system. This is very uncertain because we do not know how deeply the warming would penetrate into the oceans. If the response of the climate system involves deep ocean circulations, the climate response time may be of the order of centuries. This is currently a subject of intense study using state-of-the-art coupled atmosphere–ocean general circulation computer models.

Because of the inertia, the radiative budget of our climate system at present is not balanced. That is, the earth currently receives more solar energy (in the first term on the right-hand side of Eq. (8.24)) than it radiates back to space (in the second term in that same equation). This radiative imbalance was estimated in 2003 to be 0.85 ± 0.15 watts m^{-2} by Hansen et al. (2005) using a combination of measurements and climate model runs. The imbalance is due to the thermal inertia of our climate system. This we have modeled by the left-hand side of Eq. (8.24). Since the right-hand side of Eq. (8.24) represents the difference between the radiative input and output of the earth, the left-hand side can be estimated from this measured imbalance, yielding, for 2003 values:

$$R\frac{\partial}{\partial t}\delta T \approx 0.85 \text{ watts m}^{-2}.$$

The earth has warmed globally by $0.6 \pm 0.2°C$ from 1880 to 2003 (IPCC, 2001). The time-like quantity τ can now be assigned a value:

$$\tau \equiv R/B \approx 0.85/[(1.90)(0.6/123)] \approx 90 \pm 46 \text{ years}.$$

As can be seen in the time-dependent solution (8.28), the lag time for the climate system response is not τ; instead, it is $\sim g\tau$, which is ~ 270 years for a climate gain factor of $g \sim 3$.

It probably takes more than 200 years after the greenhouse gases have been curbed for our climate system to reach the predicted equilibrium! If the carbon dioxide is doubled and maintained at that level for 200 years, we will reach a global warming of 4°C. In the meantime, that predicted equilibrium warming is less relevant.

Asymptotic Solution for the Initial Growth Period

If this estimate of our climate inertia is correct, we are currently still in an initial growth period, with $t/g\tau$ small. Expanding in a Taylor series:

$$\exp\left\{-\frac{t}{g\tau}\right\} \cong 1 - \frac{t}{g\tau} + \frac{t^2}{2(g\tau)^2},$$

and substituting this into Eq. (8.28), we find

$$\delta T(t) \cong \frac{(1-\alpha)}{2B\tau}bt^2. \tag{8.31}$$

This is a surprising result: During the initial warming phase—we may currently still be in such a phase—the warming is approximately independent of the climate feedback factor! Take t to be the present time and $t/\tau \sim (123/90)$. The additional radiative forcing since the preindustrial period has been estimated to be $(1 - \alpha)\delta Q(t) \sim 1.80 \pm 0.85$ watts m$^{-2} = (1 - \alpha)bt$ (IPCC, 2001; Hansen et al., 2005). Using these numbers we find $\delta T(t) \sim 0.7°C$, which is close to the warming thought to have occurred during the past century, and happens to be close to the back-of-the-envelope estimate given earlier. This is not to say that the various feedback processes that increase the climate's sensitivity are unimportant. They are important in determining the eventual equilibrium warming. However, it may take a couple of centuries to reach that predicted larger equilibrium value of 4°C.

8.9 Exercises

1. *Radiative equilibrium temperature*

    Determine the radiative equilibrium temperature distribution as a function of y for the current climate, with the ice line located at $y_s = 0.95$. The radiative equilibrium solution is the solution of Eq. (8.7) with no dynamical transports.

    a. Plot such a solution. Is such a temperature distribution consistent with an ice edge located at $y = 0.95$? Why?

    b. If y_s is not fixed at the present value but is allowed to vary so that the temperature is greater than T_c to the south of the ice edge and lower to the north of the ice edge, where would such a location be?

2. *Stability of radiative equilibrium temperature*

    Determine the stability of the radiative equilibrium solution to small perturbations. Does your result apply to finite perturbations?

3. *Stabilizing effect of dynamics*

    a. Calculate how low the solar input Q must be for the onset of ice under radiative equilibrium.

    b. Do the same calculation as in (a), except now with transport C nonzero.

c. Based on the results in (a) and (b), do you think the effect of dynamical transport of heat is stabilizing or destabilizing to the climate? Why do you think this is so? How can you reconcile this result with the known destabilizing effect of dynamics when the ice line moves past the midlatitudes?

4. Unfreezing the snowball Earth

If the earth is completely covered with ice, to what must the total solar input (Q) be increased in order for the ice to melt at the equator? At that higher Q, is the partially ice-covered climate stable? What is the eventual climate at that value of solar input?

5. Sensitivity of our current climate

To measure the sensitivity of our current climate to the catastrophe of a runaway freeze, calculate the percentage change in Q needed to move the ice line from its present location of $72°$ to the unstable latitude of $34°$.

6. Diffusive dynamical transport model

A better form for the dynamical transport of heat from one latitude to the other is that of a diffusive process (see North, 1975). His model for the transport of heat is

$$D(y) = \mu a^2 \Delta T,$$

where μ is an empirical diffusion coefficient. The Laplacian operator in spherical coordinates is

$$\Delta = \frac{1}{a^2} \frac{d}{dy} (1 - y^2) \frac{d}{dy}.$$

When integrated over the globe the effect of transport should be zero. The radius of the earth is a.

a. Find the equilibrium solution $T^*(y)$ for an ice-free globe. Assume a power series solution of the form $T^* = a_0 + a_1 y + a_2 y^2 + \cdots$. (Note that powers of y higher than 2 are not needed; neither are odd powers of y.) In addition, find the consistency condition on Q such that the temperature at the pole is above that for glaciation (T_c).

b. Repeat (a) but for an ice-covered globe. In this case find the consistency condition on Q such that the temperature at the equator is lower than that for glaciation.

7. Eleven-year solar cycle

The sun's radiant output fluctuates on an 11-year periodic cycle that is modeled by $Q = Q_0 + \delta Q$, where $\delta Q(t) = a\cos(\omega t)$, with $\omega = 2\pi/(11 \text{ years})$. Solve the time-dependent equation (8.24) for the periodic temperature response of the atmosphere near the surface, $\delta T(t)$. Show that it can be written in the form

$$\delta T(t) = \frac{\delta Q(t - \Delta) \cdot (1 - \alpha)g/B}{\sqrt{1 + \epsilon^2}},$$

where $\epsilon = g\omega\tau$, and $\omega\Delta = \tan^{-1}(\epsilon)$. Δ is the time lag of the response, and the factor in the denominator gives the reduction in amplitude from the equilibrium value because of the periodic nature of the response.

8. Climate gain inferred from climate's response to the solar cycle

The variability of the sun's radiation through the 11-year solar cycle has been measured since 1979 by earth-orbiting satellites. We know that the solar constant varies by 0.06% from solar minimum to solar maximum. Referring to the parameters in exercise 7, we know that $2a/Q_0 = 0.06\%$. So $2a = 0.2$ watts per square meter. The atmosphere's temperature response is found to lag only slightly (by about 1 year) and its magnitude is measured near the surface to be about $0.2°C$ on a global average from minimum to maximum. Use these values to deduce the climate gain factor g, and show that it should be about $g \sim 3$.

9. Time-dependent global warming

We consider the scenario of a period of radiative perturbation growing with rate b, $\delta Q(t) = a\exp(bt)$ for $-\infty < t < t_s$, before a policy action to curb the growth at a future time $t = t_s$:

$$\delta Q(t) = \delta Q(t_s) \quad \text{for } t > t_s.$$

By solving Eq. (8.24), show that the atmosphere's response to this forcing, subject to the initial condition $\delta T(-\infty) = 0$, is

$$\delta T(t) = \frac{(1 - \alpha)\delta Q(t)}{B} g \frac{1}{(1 + \gamma)} \quad \text{for } t < t_s$$

and

$$\delta T(t) = \frac{(1 - \alpha)\delta Q(t_s)}{B} g \left[1 - \exp\left\{ -\frac{(t - t_s)}{g\tau} \right\} \frac{\gamma}{1 + \gamma} \right] \quad \text{for } t > t_s,$$

where $\gamma = b(R/B)g$.

10. Asymptotic limits of the global warming solution

The nature of the solution obtained in exercise 9 depends on the nondimensional quantity $\gamma = b(R/B)g = bg\tau$. This is a measure of how fast the forcing is increasing relative to the natural response time of the atmosphere–ocean system. To help understand the exact solution we next consider the solution in different asymptotic limits with respect to γ.

a. *The slow growth limit, $\gamma \ll 1$.* Show that the solution is given approximately by

$$\delta T = \frac{(1 - \alpha)\delta Q(t)}{B}g \quad \text{for } t < t_s \quad \text{and} \quad \text{for } t > t_s,$$

which is in the same form as the equilibrium solution, except with instantaneous forcing. We call this the quasi-equilibrium solution.

b. *The rapid growth limit, $\gamma \gg 1$.* Show that the solution becomes

$$\delta T(t) = \frac{(1 - \alpha)\delta Q(t)}{B} \cdot \frac{1}{b\tau} \quad \text{for } t < t_s.$$

The surprising result is that the climate's response to rapid radiative forcing is independent of the climate gain factor g.

c. *The rapid growth limit, $\gamma \gg 1$.* Show that for $t > t_s$, the time-dependent solution is independent of γ and that the time scale for approach to equilibrium is given by $g\tau$.

9

Interactions: Predator–Prey, Spraying of Pests, Carnivores in Australia

Mathematics developed:
> system of two coupled ordinary differential equations;
> equilibria; linearization about the equilibria to determine linear
> stability properties

9.1 Introduction

So far we have considered the equations governing the evolution of only one species. The problem becomes more interesting when there are two or more interacting species. The results are often unexpected (and counterintuitive). This is precisely the reason why having a mathematical model for the interactions is helpful.

The fish population in the upper Adriatic Sea forms an interesting ecological system. The sharks prey on the small fish, while the small fish feed on the plankton. They reach a balance with human fishing. Only small changes in populations were observed to occur from year to year prior to World War I. However, during World War I fishing was suspended, resulting, for a while, in more small fish than usual. Because of the war no observations were made. Soon thereafter, the population of sharks increased since they had more than the usual amount of food available. The increased number of sharks in turn devoured so many of the fish that when the fishermen returned after the war, they found, contrary to what they expected, that there was less than the usual catch of small fish. The growth of the small fish led to the growth of the sharks, which then led to the decline of the small fish.

The equations often used to model the above situation, and others like it—the "predator–prey" problem—are a coupled set of nonlinear ordinary differential equations. Again we study their equilibria and the stability of these equilibria.

9.2 The Nonlinear System and Its Linear Stability

Let $x(t)$ and $y(t)$ be the two interacting species

$$\frac{d}{dt}x = f(x, y), \tag{9.1}$$

$$\frac{d}{dt}y = g(x, y). \tag{9.2}$$

f and g are in general nonlinear functions of x and y. The procedure we use to gain some information on the possible behaviors of this system is as follows:

1. Find the equilibrium solutions x^* and y^* of the above system by solving the simultaneous algebraic equations:

$$f(x^*, y^*) = 0 \text{ and } g(x^*, y^*) = 0. \tag{9.3}$$

2. Determine if the equilibrium is stable or unstable. This can be done easily only for small perturbations from the equilibrium. To do this, we first *linearize* the nonlinear equations about the equilibrium solution (x^*, y^*). We write

$$x(t) = x^* + u(t), \tag{9.4}$$

$$y(t) = y^* + v(t), \tag{9.5}$$

$$\frac{d}{dt}x = \frac{d}{dt}(x^* + u) = \frac{d}{dt}u, \quad \frac{d}{dt}y = \frac{d}{dt}v,$$

since (x^*, y^*) do not depend on time (i.e., they are at equilibrium). Next we expand f and g about the equilibrium in a Taylor series:

$$f(x, y) = f(x^*, y^*) + \overbrace{\frac{\partial f}{\partial x}(x^*, y^*)}^{\equiv a_{11}} \cdot u + \overbrace{\frac{\partial f}{\partial y}(x^*, y^*)}^{\equiv a_{12}} \cdot v$$

$$+ \text{ terms involving } u^2, v^2, uv, u^3, v^3, \text{ etc.}$$

$$\cong a_{11}u + a_{12}v, \tag{9.6}$$

since $f(x^*, y^*) = 0$ from Eq. (9.3). The process of dropping the higher order terms (the nonlinear terms) is called *linearization*.

It is valid if we are to study only the behavior of the solution very near the equilibrium points (x^*, y^*) (so that u and v are small in some measure, and u^2, v^2, uv, etc., are smaller still). Similarly

$$g(x, y) \cong a_{21}u + a_{22}v, \tag{9.7}$$

where

$$a_{21} = \frac{\partial g}{\partial x}(x^*, y^*) \text{ and } a_{22} = \frac{\partial g}{\partial y}(x^*, y^*).$$

3. We end up, after this process of linearization, with a coupled *linear* system:

$$\frac{du}{dt} = a_{11}u + a_{12}v,$$

$$\frac{dv}{dt} = a_{21}u + a_{22}v, \tag{9.8}$$

or in matrix form:

$$\frac{d}{dt}\begin{bmatrix} u \\ v \end{bmatrix} = \begin{bmatrix} a_{11} & a_{12} \\ a_{21} & a_{22} \end{bmatrix}\begin{bmatrix} u \\ v \end{bmatrix},$$

which is sometimes written in the following notation:

$$\frac{d}{dt}\mathbf{u} = A\mathbf{u}.$$

4. For linear equations with constant coefficients, we try exponential solutions:

$$u(t) = u_0 e^{\lambda t}, \; v(t) = v_0 e^{\lambda t}.$$

Substituting into the ordinary differential equation, we get

$$\lambda u_0 = a_{11}u_0 + a_{12}v_0,$$

$$\lambda v_0 = a_{21}u_0 + a_{22}v_0,$$

or

$$\begin{bmatrix} a_{11} - \lambda & a_{12} \\ a_{21} & a_{22} - \lambda \end{bmatrix}\begin{bmatrix} u_0 \\ v_0 \end{bmatrix} = 0.$$

To have nontrivial solutions, we must have

$$\det \begin{bmatrix} a_{11} - \lambda & a_{12} \\ a_{21} & a_{22} - \lambda \end{bmatrix} = 0. \tag{9.9}$$

Expanding out the determinant we obtain the characteristic equation

$$\lambda^2 - (a_{11} + a_{22})\lambda + (a_{11}a_{22} - a_{12}a_{21}) = 0,$$

which we rewrite as

$$\lambda^2 - p\lambda + q = 0, \tag{9.10}$$

where

$$p \equiv a_{11} + a_{22} = T_r\{A\},$$

$$q \equiv a_{11}a_{22} - a_{12}a_{21} = \det A,$$

are the trace and determinant of the matrix A. Solving the quadratic equation we obtain the two eigenvalues

$$\lambda_1 = \frac{p}{2} + \frac{\sqrt{p^2 - 4q}}{2}, \quad \lambda_2 = \frac{p}{2} - \frac{\sqrt{p^2 - 4q}}{2}. \tag{9.11}$$

The two parameters p and q determine the stability of the system under consideration. Refer to Figure 9.1 for the following cases.

a. If $q < 0$, the two eigenvalues are real, one positive and one negative. In this case, the equilibrium ($u = 0$, $v = 0$) is a *saddle point*, which is unstable. ($e^{\lambda_1 t}$ grows exponentially; $e^{\lambda_2 t}$ decays exponentially. The general solution is a linear combination of $e^{\lambda_1 t}$ and $e^{\lambda_2 t}$, and hence unstable.)

b. If $0 < q < p^2/4$, we will have two real roots of the same sign. For $p < 0$, both roots are negative and we will have a *stable node*. For $p > 0$, both roots are positive and we have an *unstable node*.

c. If $q > p^2/4$, both roots are complex and we will have oscillations. The amplitude of the oscillation will increase or decrease in t depending on the sign of p. For $p < 0$, we have a *stable focus*. For $p > 0$, we have an *unstable focus*. Finally, if $p = 0$, we have a *center* (which is neutral: borderline stability). We will explain the meaning of these terms in a moment. The

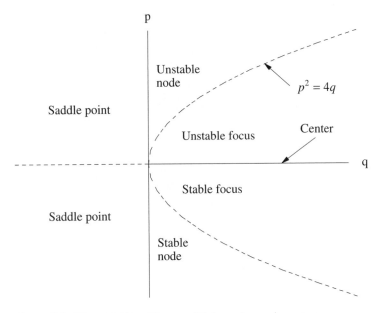

Figure 9.1. The stability of the equilibrium depends on two parameters, p and q. (Modified from original figure in Kot [2001], by permission.)

solution to Eq. (9.8) is, in general,

$$\mathbf{u} = \mathbf{u}_0^{(1)} e^{\lambda_1 t} + \mathbf{u}_0^{(2)} e^{\lambda_2 t}. \qquad (9.12)$$

$\mathbf{u}_0^{(1)}, \mathbf{u}_0^{(2)}$ are constant vectors. $\mathbf{u}_0^{(1)}$ is also called the *eigenvector* corresponding to the eigenvalue λ_1, etc.

The general solution (9.12) will grow in time (hence the origin $u = 0$, $v = 0$, is *unstable*) if either λ_1 or λ_2 has a positive real part. On the other hand, the origin is *stable* only if both λ_1 and λ_2 have a negative real part.

We shall use arrows to indicate the "trajectories" of the solutions as time increases.

Project any initial point $\mathbf{u}(0)$ onto $\mathbf{u}_0^{(1)}$ and $\mathbf{u}_0^{(2)}$. As time increases, the $\mathbf{u}_0^{(1)}$ part becomes $\mathbf{u}_0^{(1)} e^{\lambda_1 t}$; i.e., it contracts or expands by the factor $e^{\lambda_1 t}$. The $\mathbf{u}_0^{(2)}$ part will evolve according to $\mathbf{u}_0^{(2)} e^{\lambda_2 t}$.

9.3 Lotka–Volterra Predator–Prey Model

Let us now return to the problem posed in the introduction.

Fishing activity was greatly curtailed in the upper Adriatic during World War I. D'Ancona, an Italian marine biologist, showed that this

Figure 9.2. Vito Volterra (1840–1940).

period coincided with an increase in the number of sharks relative to their prey, the small fish, which eat algae. When the fishermen returned to the sea at the end of the war, they found that the fish stock had diminished severely, even though there had not been much fishing in the preceding years. At the time, Umberto was engaged to Luisa Volterra. He brought this problem to the attention of Vito Volterra, his future father-in-law and a famous mathematician (Figure 9.2).

The equations Vito wrote down to describe this situation are now known as the Lotka–Volterra equations:

$$\frac{d}{dt}x = rx - axy, \qquad (9.13)$$

$$\frac{d}{dt}y = bxy - ky. \qquad (9.14)$$

Here $x(t)$ denotes the population density of the prey, which are the small fish in this case. $y(t)$ is the population density of the predator, the sharks. The fish eat algae, which is abundant, and grow at a per capita rate $\left(\frac{1}{x}\frac{d}{dt}x\right)$ of r. The small fish are eaten by the sharks, and so their population density decreases at a per capita rate that is proportional to y, the population density of the predator, with proportionality a. The predators, on the other hand, will die off without food. So if $x = 0$, $\frac{1}{y}\frac{d}{dt}y$ decreases at the rate k, which is inversely proportional to the time it takes for the predators to die of starvation. In the presence of prey, the population of the predator grows at a per capita rate of bx, proportional to the amount of food available.

Linear Analysis

To analyze this system, we first look for the equilibria (x^*, y^*) by setting the time derivatives in Eqs. (9.13) and (9.14) to zero:

$$rx^* - ax^*y^* = 0, \quad bx^*y^* - ky^* = 0. \tag{9.15}$$

The first algebraic equation has two roots:

$$x^* = 0 \quad \text{and} \quad y^* = r/a.$$

Substituting $x^* = 0$ into the second algebraic equation, we find that we must have $y^* = 0$. Similarly, substituting $y^* = r/a$ into the second equation (or, for that matter, any nonzero value of y^*) implies that $x^* = k/b$. Thus we have found two equilibria. They are

$$(x_1^*, y_1^*) = (0, 0)$$

and

$$(x_2^*, y_2^*) = (k/b, r/a).$$

Each of these two values solves Eq. (9.15) simultaneously.

We now consider the stability of each of the two equilibria by perturbing them slightly by the amount (u, v):

$$x(t) = x^* + u(t), \quad y(t) = y^* + v(t). \tag{9.16}$$

The perturbations satisfy Eq. (9.8) and we can solve them using the method outlined in section 9.2. Alternatively the same result can be obtained in the following more intuitive way. For $(x^*, y^*) = (x_1^*, y_1^*) = (0, 0)$, the equations governing the evolution of the perturbations

(u, v) are obtainable by substituting (9.16) into Eqs. (9.13) and (9.14), yielding

$$\frac{d}{dt}u = ru - auv,$$

$$\frac{d}{dt}v = buv - kv.$$

For small perturbations, we drop the quadratic terms in favor of linear terms, giving

$$\frac{d}{dt}u \simeq ru,$$

$$\frac{d}{dt}v \simeq -kv.$$

(9.17)

The same two equations can also be arrived at using the formal procedure in (9.6) and (9.7) (In that notation we have $a_{11} = r, a_{12} = 0, a_{21} = 0$, and $a_{22} = -k$.) The solution to Eq. (9.17) is unstable (and we say the equilibrium $(x_1^*, y_1^*) = (0, 0)$ is an unstable saddle; check by evaluating p and q).

It turns out that the two equations in (9.18) are decoupled and can be solved easily to yield

$$u(t) = u(0)e^{rt}, \quad v(t) = v(0)e^{-kt}.$$

A small increase from $(0, 0)$ will lead to an exponential growth in the prey (because the predators are so few in number and algae are plentiful). A small increase in predators, on the other hand, will not lead to a growth of the predator population. Instead the predators will die of starvation because there are so few fish to prey upon for food. Nevertheless, the equilibrium $(0, 0)$ is still unstable because one of the populations does not stay low when perturbed.

Near the second equilibrium, we substitute

$$x(t) = x_2^* + u(t) = k/b + u(t),$$

$$y(t) = y_2^* + v(t) = r/a + v(t),$$

into Eqs. (9.13) and (9.14) and find

$$\frac{d}{dt}u = r\left(\frac{k}{b}+u\right) - a\left(\frac{k}{b}+u\right)\left(\frac{r}{a}+v\right)$$

$$= ru - ru - a\left(\frac{k}{b}\right)\cdot v - auv$$

$$\cong -a\left(\frac{k}{b}\right)v, \tag{9.18}$$

after ignoring the quadratic term (uv).

Similarly,

$$\frac{d}{dt}v = b\left(\frac{k}{b}+u\right)\left(\frac{r}{a}+v\right) - k\left(\frac{r}{a}+v\right)$$

$$= b\left(\frac{r}{a}\right)u + kv + buv - kv$$

$$\cong b\left(\frac{r}{a}\right)u. \tag{9.19}$$

(You can verify that the same two equations can be obtained by using the a_{ij}'s. In this case $a_{11} = 0$, $a_{12} = -a(k/b)$, $a_{21} = b(r/a)$, $a_{22} = 0$. Since $p \equiv a_{11} + a_{22} = 0$, $q \equiv a_{11}a_{22} - a_{12}a_{21} = rk > 0$, the equilibrium $(x_2^*, y_2^*) = (k/b, r/a)$ is of borderline stability, and is in fact a *center*.)

Equations (9.18) and (9.19) can in fact be solved by differentiating one of the equations and substituting the second equation to eliminate one of the unknowns:

$$\frac{d^2}{dt^2}u = -a(k/b)\frac{d}{dt}v = -kru.$$

This equation, when written in the form

$$\frac{d^2}{dt^2}u + kru = 0,$$

is recognized as the equation for a harmonic oscillator. The solution is

$$u(t) = A\cos(\sqrt{kr}\,t) + B\sin(\sqrt{kr}\,t). \tag{9.20}$$

From (9.18), $v = -b/(ka)\frac{d}{dt}u$, and we have

$$v(t) = (b/a)\sqrt{r/k}[A\sin\sqrt{kr}t - B\cos\sqrt{kr}t]. \tag{9.21}$$

It is therefore seen that the solution is oscillatory and is periodic with period $2\pi/\sqrt{kr}$. It can be shown that the prey perturbation leads the predator perturbation by a quarter cycle (or, equivalently, lags behind it by three quarter cycles). (Solution (9.20) can be rewritten, without loss of generality, as

$$u(t) = C\cos(\sqrt{kr}t + \delta),$$

where C and δ are arbitrary constants. Solution (9.21) can then be rewritten as

$$v(t) = (b/a)\sqrt{r/k} \cdot C\sin(\sqrt{kr}t + \delta)$$

$$= (b/a)\sqrt{r/k} \cdot C\cos(\sqrt{kr}t + \delta - \pi/2).)$$

Nonlinear Analysis

From the above analysis, we conclude that the nonlinear system, (9.13) and (9.14), does not possess any equilibria that are stable. So we don't expect that the solution to the system will approach equilibrium as a steady state. We further showed that the solution in the neighborhood of $(x^*, y^*) = (k/b, r/a)$ is periodic. The question remains as to the behavior of the solution at finite distances from this "equilibrium." We will next show that the solution remains periodic.

Figure 9.3 displays the direction field of the solution at each point (x, y) of the phase plane. The slope of the direction field is obtained when one divides (9.14) by (9.13):

$$\frac{dy}{dx} = \frac{dy/dt}{dx/dt} = \frac{y(bx - k)}{x(r - ay)}. \tag{9.22}$$

At each point (x, y), the slope dy/dx can be found by evaluating the right-hand side and plotted as a short line. The result is plotted in Figure 9.3. The direction of the arrows in the figure indicates the direction of increase as t increases. This can be inferred from (9.13) and (9.14) for any value of (x, y). In particular, the arrows should point towards the right for $y < r/a$ and towards the left for $y > r/a$. They should point up for $x > k/b$ and down for $x < k/b$. Figure 9.3 shows that the nonlinear solution cycles around the point (x_2^*, y_2^*). A cyclic solution, as indicated by closed loops in the phase plane, is a periodic solution.

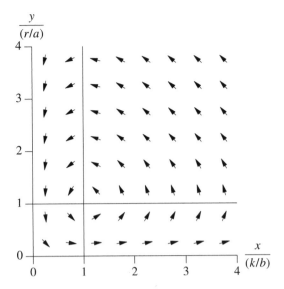

Figure 9.3. The direction field of Lotka–Volterra equations. (Modified from original figure in Kot [2001], by permission.)

Since the solution is periodic, one can define

$$x_{\text{average}} = \frac{1}{T} \int_{t_0}^{t_0+T} x(t)\, dt,$$

$$y_{\text{average}} = \frac{1}{T} \int_{t_0}^{t_0+T} y(t)\, dt,$$

as the value averaged over a period, where T is the period of the oscillation, and t_0 is any time. Volterra showed that these averaged values can be found exactly by integrating the original equations (9.13) and (9.14): from (9.13) one has

$$\frac{1}{x}\frac{d}{dt}x = r - ay.$$

Integrating both sides with respect to t yields, for the left-hand side:

$$\int_{t_0}^{t_0+T} \frac{1}{x}\frac{dx}{dt}\, dt = \ln\left[\frac{x(t_0 + T)}{x(t_0)}\right] = \ln\left[\frac{x(t_0)}{x(t_0)}\right] = 0,$$

and for the right-hand side:

$$\int_{t_0}^{t_0+T} (r - ay)dt = rT - aTy_{\text{average}}.$$

Thus, we have

$$0 = rT - aTy_{\text{average}},$$

yielding

$$y_{\text{average}} = r/a. \tag{9.23a}$$

A similar consideration for (9.14) will yield

$$x_{\text{average}} = k/b. \tag{9.23b}$$

Note that these are the same as the "equilibrium" (x_2^*, y_2^*), about which the solution is cycling.

Some rather counterintuitive results now follow. If the growth rate r of the prey is somehow increased, we don't get a corresponding increase in the average prey population (because x_{average} is independent of r). Instead, the predator population increases. If the death rate of the predator is increased, the prey's average population is increased, while the average population of the predator is unaffected.

9.4 Harvesting of Predator and Prey

Let us now return to modeling the problem that concerned D'Ancona originally, and add a harvesting term to both the prey and predator equations (i.e., the harvesting is indiscriminate with respect to fish and shark):

$$\frac{dx}{dt} = rx - axy - qEx,$$

$$\frac{dy}{dt} = bxy - ky - qEy.$$

These can be rewritten as

$$\frac{dx}{dt} = (r - qE)x - axy, \tag{9.24}$$

$$\frac{dy}{dt} = bxy - (k + qE)y. \tag{9.25}$$

Equation (9.24) is the same as (9.13) (without harvesting), except with the prey growth rate r reduced to $r - qE$. Equation (9.25) is the same as (9.14) except with the predator death rate k increased to $k + qE$. The average prey and predator population densities then become, by modifying (9.23):

$$x_{average} = (k + qE)/b, \tag{9.26}$$

$$y_{average} = (r - qE)/a. \tag{9.27}$$

During the war, when the fishing effort (E) is reduced, it actually increases the average shark population while reducing the average fish population, as (9.26) and (9.27) show.

What happens after the war, when fishing is increased?

Indiscriminate Spraying of Insects

The cottony cushion scale insect (*Icerya purchasi*) was accidentally introduced into America in 1868 from Australia and threatened to destroy the American citrus industry. To control this insect, its natural Australian predator, a ladybird beetle (*Novius cardinalis*), was introduced, and the beetles reduced the scale insects to a low level. When DDT was discovered and found to kill the scale insects, it was applied by the orchardists in the hope of further reducing the scale insects. The spraying of DDT killed insects and beetles indiscriminately at the same per capita rate. Discuss the effect of spraying. Is it beneficial to the citrus industry?

9.5 The Case of the Missing Large Mammalian Carnivores

Meganesia, which comprises Australia, Tasmania, and New Guinea, is unusual in its extraordinary lack of large mammalian carnivores throughout its history. One possible cause may be the notoriously infertile soil of this old continent with a stable geological history. It takes an area of grassland with billions of individual grasses to sustain a few thousand large herbivores. These, in turn, may be able to support fewer than one hundred large carnivores. If the environment is poor, large herbivores will be rare and thinly spread. The density of the prey is so low that a population of large mammalian meat-eaters cannot be sustained. Cold-blooded reptiles, on the other hand, eat far less than mammals do, having no need to create inner body heat. This scenario may explain what happened in Meganesia, home to a remarkable array of carnivorous reptiles (Flannery, 1993).

Consider the predator–prey equations:

$$\frac{dx}{dt} = rx(1 - x/K) - axy, \tag{9.28}$$

$$\frac{dy}{dt} = -ky + bxy, \tag{9.29}$$

where y is the population of meat-eating predator. If the prey x are the herbivores, whose population growth follows the logistic equation (resource limited), with K being the carrying capacity of the herbivores that the vegetation can support in the absence of the predator. The parameter k/b measures how efficiently the predator utilizes food to make babies. The more efficient predators, such as large reptilian carnivores, have smaller values of k/b.

a. When the vegetation is poor, the carrying capacity K for the herbivore prey is low. If the predator is a large mammalian carnivore with high k/b, specifically,

$$K < k/b,$$

show that the predator will become extinct. (Find the equilibria and determine their stability.)

b. Replace the predator by large reptilian carnivores with

$$k/b < K.$$

Show that there exists a stable equilibrium with a positive predator population.

Solution

$\frac{dx}{dt} = f(x, y), \frac{dy}{dt} = g(x, y)$:

$$f(x, y) = x[r(1 - x/K) - ay],$$

$$g(x, y) = y[-k + bx].$$

The equilibria are given by either

$$\text{(i) } x^* = 0, \ y^* = 0,$$

$$\text{(ii) } x^* = K, \ y^* = 0, \text{ or}$$

$$\text{(iii) } x^* = k/b, \ y^* = (r/a)(1 - k/(bK)).$$

a. For the case of large mammalian predators,

$$k/(bK) > 1,$$

(iii) is not feasible because it would have required a negative predator population. Of the remaining two equilibria, (i) is unstable and (ii) is stable. Thus the only stable long-term solution is a population of herbivores at its carrying capacity, K, with no predator population ($y^* = 0$). This population of herbivores is unable to sustain any large carnivores.

b. For the case of reptilian predators,

$$k/(bK) < 1,$$

(iii) is a possibility (with $y^* > 0$). For this case (ii) becomes unstable and (iii) is stable. Thus it is possible for predator and prey to coexist.

Stability analysis
For both cases (a) and (b), the equations linearized about (i) are

$$\frac{du}{dt} = ru, \quad \frac{dv}{dt} = -kv,$$

where

$$u = x - x^*, \quad v = y - y^*, \quad \text{with } x^* = 0, \ y^* = 0.$$

Thus (i) is an unstable saddle because $u(t) = u(0)e^{rt}$ and grows exponentially, although $v(t) = v(0)e^{-kt}$ decays.

For case (a), the linearized equations about (ii) are

$$\frac{du}{dt} = -ru - aKv, \quad \frac{dv}{dt} = -(k - bK)v,$$

where $u = x - x^*, v = y - y^*$, with $x^* = K, y^* = 0$. So

$$v(t) = v(0)e^{-(k-bK)t} \quad \text{decays because } k - bK > 0,$$

and

$$u(t) = -\frac{aK\,v(0)}{(r - k + bK)}e^{-(k-bK)t} - Ce^{-rt}$$

also decays for $k - bK > 0$. Thus the equilibrium (ii) is stable for case (a).

For case (b), $k - bK < 0$, $u(t)$, and $v(t)$ grow exponentially, and so (ii) is unstable.

The stability of (iii) is to be determined by linearizing the equations about (iii):

$$\frac{du}{dt} = -\frac{kr}{bK}u - \frac{ka}{b}v, \quad \frac{dv}{dt} = (r/a)(1 - k/bK) \cdot bu.$$

Combining, we get

$$\frac{d^2u}{dt^2} + \frac{kr}{bK}\frac{du}{dt} + kr(1 - k/bK)u = 0.$$

Assume an exponential solution

$$u(t) = u(0)e^{\lambda t}$$

yields

$$\lambda^2 + \frac{kr}{bK}\lambda + kr(1 - k/bK) = 0;$$

then

$$\lambda = -\frac{kr}{2bK} \pm \left\{\frac{1}{4}\left(\frac{kr}{bK}\right)^2 - kr(1 - k/bK)\right\}^{1/2}.$$

Since the real part of λ is always negative, the equilibirum (iii) is stable for case (b).

9.6 Comment

The mathematical model appears to have confirmed the conjecture by Flannery (1993) and neatly explained why Australia lacks large meat-eating mammals. However, reality might not be as tidy.

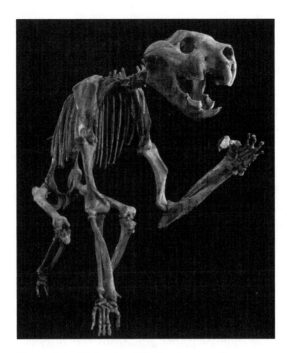

Figure 9.4. Skeleton of *Thylacoleo carnifex*.

Flannery mentioned that there were approximately 60 species of mammals that weighed more than 20 lbs. before the arrival of humans in Australia some 50,000 years ago, but he thought all except three were herbivores and there were no large meat-eaters among them.

The Tasmanian devil, weighing less than 20 lbs., is best described as a miniature marsupial hyena. The spotted-tail quall is the marsupial version of the weasel. Both still survive. When the extinct marsupial lion (*Thylacoleo carnifex*) was first described from fossils in the 1850s, it was thought to have been among the "fellest and most destructive of predatory beasts," the ecological equivalent of a lion. Since then its size has been revised steadily downward. Flannery thought it was the marsupial equivalent of a medium-sized cat on other continents. Using the only known complete skeleton of *Thylacoleo carnifex* (Figure 9.4), a specimen from Moree, a team of paleontologists headed by Stephen Wroe from the University of New South Wales just recently revised its weight dramatically upward, to as much as 164 kg, with an average of between 100 kg and 130 kg. "That would be a good-size female lion or tiger," Wroe said. "It has been suggested that big mammalian predators could not exist in Australia because there wasn't enough food for them," he explained. "But these new measurements show that *Thylacoleo* was a big kick-arse carnivore by any standard, blowing the old theory right out of the water" (*News in Science*, April 6, 1999).

9.7 More Examples of Interactions

Example 1. *Romantic Romeo, Fickle Juliet* (Strogatz, 1988)
R: Romeo's love for Juliet.

$$\frac{dR}{dt} = aJ, \ a > 0:$$

Romeo's love grows in response to Juliet's love.
J: Juliet's love for Romeo.

$$\frac{dJ}{dt} = -bR, \ b > 0:$$

Juliet's love decreases in response to Romeo's love for her. Can they be simultaneously in love with each other?

This system of linear equations has only one equilibrium, at the origin ($R = 0, J = 0$). The stability of this equilibrium can be determined by the eigenvalue method outlined earlier. However, since this system is linear, we can actually obtain an exact solution, not just information about the stability of the point ($R = 0, J = 0$). We divide the two equations:

$$\frac{dR}{dJ} = \frac{dR/dt}{dJ/dt} = \frac{aJ}{-bR},$$

$$bR\,dR + aJ\,dJ = 0,$$

$$bR^2(t) + aJ^2(t) = \text{constant}, \quad C = bR(0)^2 + aJ^2(0).$$

The orbits are ellipses in the R-J plane. Romeo and Juliet can be simultaneously in love with each other only one quarter of the time.

The detailed time evolution of the solution can also be found:

$$\frac{d^2R}{dt^2} = \frac{d}{dt}\frac{d}{dt}R = \frac{d}{dt}(aJ) = a\frac{dJ}{dt} = a(-bR),$$

so

$$\frac{d^2R}{dt^2} + abR = 0.$$

J

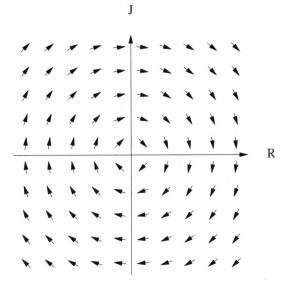

R

Figure 9.5. The direction field of the Romeo–Juliet system.

If we try an exponential solution of the form

$$R(t) = ce^{\lambda t},$$

we will find that $\lambda^2 = -ab$. So $\lambda_1 = \sqrt{ab}i$, $\lambda_2 = -\sqrt{ab}i$.

The general solution is

$$R(t) = c_1 e^{i\sqrt{ab}t} + c_2 e^{-i\sqrt{ab}t},$$

$$\left(J = \frac{dR}{dt}/a = \frac{i\sqrt{ab}}{a}c_1 e^{i\sqrt{ab}t} - \frac{i\sqrt{ab}}{a}c_2 e^{-i\sqrt{ab}t} \right).$$

Alternatively, since $e^{i\theta} = \cos\theta + i\sin\theta$ (Euler's formula), we can rewrite the solution as

$$R(t) = c_3 \cos(\sqrt{ab}t) + c_4 \sin(\sqrt{ab}t),$$

where the c's are arbitrary constants:

$$J = -\sqrt{\frac{b}{a}}c_3 \sin(\sqrt{ab}t) + \sqrt{\frac{b}{a}}c_4 \cos(\sqrt{ab}t).$$

To see the phase relationship between R and J, consider the initial special condition $R(0) = 0$. Then $c_3 = 0$. In this case

$$R(t) = c_4 \sin(\sqrt{ab}\, t),$$

$$J(t) = \sqrt{\frac{b}{a}} c_4 \cos(\sqrt{ab}\, t) = \sqrt{\frac{b}{a}} c_4 \sin\left(\sqrt{ab}\, t + \frac{\pi}{2}\right)$$

$$= \sqrt{\frac{b}{a}} R\left(t + \frac{\pi}{2} / \sqrt{ab}\right).$$

Juliet's feeling lags Romeo's by a quarter cycle. (See Figure 9.5.)

Example 2. *The CONCOM Model*
The CONCOM Model, the conventional combat model, is one special type of Lanchester Models for Combat or Attrition (Lanchester, 1914).

In a Lanchester model, an x-force and a y-force are engaged in a battle of attrition. The variables $x(t)$ and $y(t)$ denote the strength of the forces at time t. A Lanchester model assumes that, for force x:

$$\frac{dx}{dt} = -(OLR + CLR) + RR,$$

where OLR is the operational loss rate, the loss rate due to diseases, desertions, and other noncombat mishaps. CLR is the combat loss rate, and RR is the reinforcement rate. A similar equation applies to y. The CONCOM model is

$$\frac{dx}{dt} = -ax - by + P(t),$$

$$\frac{dy}{dt} = -cx - dy + Q(t).$$

The operational loss rates are assumed to be proportional to the number of one's own troops. The combat loss rate is modeled differently depending on the type of warfare being conducted. For conventional warfare, every member of a conventional force is within range of the enemy and that conventional force x has a loss rate by; that is, proportional to the enemy strength but independent of x, its own number. x's loss rate is the same whether $x = 10{,}000$ or $1{,}000$. The limiting factor is how many shots y can fire per minute (think U.S. Civil War).

$P(t)$ is the rate of reinforcement for x.

(The situation is different for ancient warfare, where CLR for x should be proportional to the amount of x-force in one-on-one combat with y.

For this case, one should model the per capita loss rate $(-\frac{1}{x}\frac{d}{dt}x)$ due to combat loss as by, so $CLR = bxy$. The same reasoning should apply to the loss rate of fish eaten by sharks. It is similar to that of one-on-one combat. So one should model the per capita loss rate of the fish as proportional to the number of sharks.)

A simplified CONCOM model is one without operational loss and replacement:

$$\frac{dx}{dt} = -by,$$

$$\frac{dy}{dt} = -cx.$$

The attrition rate of each belligerent is proportional to the size of the adversary. As before, we divide the two equations:

$$\frac{dy}{dx} = \frac{dy/dt}{dx/dt} = \frac{cx}{by},$$

$$by\,dy = cx\,dx.$$

Integrating,

$$b[y(t)^2 - y_0^2] = c[x(t)^2 - x_0^2].$$

Another way of expressing the solution is as

$$by(t)^2 - cx(t)^2 = by_0^2 - cx_0^2 \equiv K \text{ a constant.}$$

The orbits are hyperbolas in the x–y plane. See Figure 9.6. We consider that x wins the war of attrition if y vanishes first. Thus x wins if $K < 0$ and y wins if $K > 0$. A stalemate occurs if $K = 0$, which is

$$by_0^2 = cx_0^2$$

or

$$y_0 = \sqrt{\frac{c}{b}}x_0, \quad \frac{c}{b} = \left(\frac{y_0}{x_0}\right)^2.$$

It says that to stalemate an adversary three times as numerous, it does not suffice to be three times as effective; you must be nine times as effective! This is referred to as the square law.

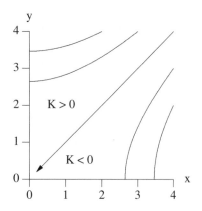

Figure 9.6. The square law of the CONCOM combat model.

9.8 Exercises

1. Arms races

Let $x(t)$ and $y(t)$ be the "war potential" of nations A and B. We can measure the potential in terms of the level of armaments of each country. The following system of equations models $x(t)$ and $y(t)$:

$$\frac{dx}{dt} = r(x_0 - x) + ay,$$

$$\frac{dy}{dt} = r(y_0 - y) + bx,$$

where $r = 1/(5 \text{ years})$ gives the rate at which a country is capable of change and is measured by the lifetime of a parliament. a and b measure how quickly a nation can arm itself in response to external threats. $a = b = 1/(1 \text{ years})$.

a. In the absence of foreign threats ($a = 0, b = 0$), each nation has an underlying equilibrium war potential. Find that equilibrium and determine its stability.

b. In the presence of interactions between the two nations, there is a buildup of war potential in response to the enemy's armament. Find the new equilibrium and determine the condition for stability (peace). The condition should be in terms of r, a, and b.

c. For the European arms race of 1909–1914, France was allied with Russia (A), and Germany with Austria-Hungary (B). Would the arms race lead to war? Use the numerical values given earlier for r, a, and b.

2. *Romeo and Juliet*

Let $R(t)$ be the love of Romeo for Juliet and let $J(t)$ be the love of Juliet for Romeo. (Let's not worry about how we measure "love".) Romeo's love for Juliet grows in response to Juliet's love (with rate b) for him and vice versa. The lovers, however, are rather cautious; they tried to rein in their love for each other (with rate a). Thus the governing equations are

$$\frac{d}{dt}R = -aR + bJ,$$

$$\frac{d}{dt}J = -aJ + bR.$$

Discuss what happens to their love if their caution a is larger than their responsiveness b, and if a is smaller than b. You can either use the method of equilibrium and its stability to support your conclusion, or you can solve these linear equations exactly.

3. *Guerilla combat (GUERCOM)*

Two guerilla forces, with troop strengths $x(t)$ and $y(t)$, are in combat with each other without reinforcement. Suppose the territory is rather large and full of places to hide. The y-force needs to find the x-force first before it can inflict combat losses, and the higher the x, the easier it is for them to be found. Therefore, the combat loss rate for the x-force should be proportional $x \cdot y$. (This is unlike conventional warfare, where the full force of x is open to be shot at by y, and so the combat loss rate for x shouldn't depend on the total number of the x-force.) Thus

$$\frac{dx}{dt} = -axy,$$

$$\frac{dy}{dt} = -bxy,$$

where a is the combat effectiveness of the y-force and b is that of the x-force. Suppose initially that x_0, y_0 are the troop strengths for the x- and y-forces, and that x_0 is three times as numerous as y_0. How much more effective must the y-force be to stalemate its enemy?

4. *Guerilla–conventional combat (VIETNAM)*

In Vietnam, a conventional force (the United States) opposed a guerilla force (the Vietcong). This conventional force $y(t)$ is out in the open to be shot at; its combat loss rate is limited only by the number of the x-force that can shoot. So the model for VIETNAM is

$$\frac{dx}{dt} = -axy,$$

$$\frac{dy}{dt} = -bx.$$

The combat effectiveness of y against x can be measured by the constant a in x's combat loss rate. a is proportional to the ratio between the exposed area of the body of a single guerilla combatant, A_g (~ 2 sq. ft.), and the area A_x occupied by the guerilla force. So

$$a = c_1 \frac{A_g}{A_x}.$$

If each guerilla combatant ranges over 1,000 sq. ft. and the force is spread out, then $A_x = (1,000 \text{ sq. ft.}) \times x_0$. The constant b, which measures the combat effectiveness of the x-force against the conventional y-force, is

$$b = c_2 p_x,$$

where $p_x \cong 0.1$ is the probability that a shot by a guerilla kills an opponent. c_1 and c_2 are firing rates, and we assume that they are comparable (i.e., $c_1 \sim c_2$).

a. Derive the condition for stalemate (parabolic law).

b. Estimate the ratio of the initial forces, y_0/x_0, for the y-force to prevail.

In Vietnam, the U.S. force never exceeded its opponent by a ratio of 6. Could the United States have prevailed?

5. *Extinction of Neanderthals*

Neanderthals were the original inhabitants of Europe and their species was very stable for more than 60,000 years. Forty thousand years ago the Neanderthals were replaced by our ancestors, the early humans (Cro-Magnon), who came to Europe from Africa much later. There is evidence that the two species coexisted in some parts of Europe. The mass extinction of the Neanderthals was rapid, in 5,000 to 10,000 years.

Theories for the demise of the Neanderthals include genocide by the early humans and competition for resources with the humans. We shall investigate the latter possibility.

Let $N(t)$ be the total population of humanoids, which consists of a population of Neanderthals, $x(t)$, and early men, $y(t)$:

$$N(t) = x(t) + y(t).$$

Suppose that they lived in the same resource-limited environment and therefore the total population satisfies the logistic equation:

$$\frac{dN}{dt} = rN(1 - N/K) - \beta N, \tag{9.30}$$

where K is the total carrying capacity for all the humanoids combined, and β is their mortality rate. (We could have included β in the definition of r (and K), but we chose not to do so for convenience.) We assume $r > \beta > 0$ because the net growth rate should be positive for small population densities.

a. Suppose there is no difference in their survival skills. Write down two coupled equations for $x(t)$ and $y(t)$, in the form

$$\frac{1}{x}\frac{d}{dt}x = F(x, y) - \beta, \tag{9.31}$$

$$\frac{1}{y}\frac{d}{dt}y = F(x, y) - \beta, \tag{9.32}$$

where $F(x, y)$ is the same for x and for y. What is $F(x, y)$? You should be guided by the requirement that the sum of (9.31) and (9.32) should give you (9.30), and that if you switch x and y in (9.31) you will get (9.32).

b. Suppose the early humans are slightly better adapted to survival than Neanderthals, but the difference is tiny. Replace the human equation (9.32) by

$$\frac{1}{y}\frac{d}{dt}y = F(x, y) - (1 - \epsilon)\beta, \tag{9.33}$$

where $0 < \epsilon \ll 1$ is the mortality difference. The Neanderthal equation remains as (9.31). Find the equilibria of (9.31) and (9.33).

   c. Determine the stability of the equilibria.

   d. Discuss the implications of the results on equilibria and their stability. Is the extinction of the Neanderthals inevitable?

   e. By forming an equation for

$$\frac{d}{dt}(x/y) = \frac{1}{y}\frac{dx}{dt} - \left(\frac{x}{y}\right)\frac{1}{y}\frac{d}{dt}y = \cdots,$$

   show that $x(t)/y(t) = A_0 e^{-\epsilon\beta t}$. Suppose we measure β by the reciprocal of the lifetime of an individual, 30 years. We know from paleontological data that it took 5,000 to 10,000 years (take it to be 10,000 years) for the Neanderthals to become extinct. Take this as the time for x/y to decrease by a factor e. Estimate the mortality difference.

6. *Open-access fishery*

   In an open-access fishery, fishermen are free to come and go as they please. The fishing effort E is determined by the opportunity to make a profit. Let c be the cost of operation, and p the price the fishermen get for their catch H. Profit is given by

$$P = pH - cE,$$

   where

$$H = qEN$$

   is the harvest rate, which is proportional to the amount of fish N there are in the fishery and the effort E expended to catch them. When there is profit to be made, the fishermen would increase their effort in the hope of making even more profit. Thus

$$\frac{d}{dt}E = aP(t), \tag{9.34}$$

   where a is a proportionality constant.
   The equation governing the fish population is

$$\frac{d}{dt}N = rN\left(1 - \frac{N}{K}\right) - H. \tag{9.35}$$

   Determine the equilibria (N^*, E^*) and their stability of the coupled system (9.34) and (9.35). (Note that the only two unknowns are $N(t)$ and $E(t)$.) Assume $c/(pqK) < 1$.

7. *Epidemiology, SIR model*

A small group of infected individuals is introduced into a large (fixed) population N, who are susceptible to this contagious disease (e.g., smallpox). The number of infected, $I(t)$, increases at a rate $\frac{dI}{dt}$ that is proportional to the product of the number of infected and the number of susceptibles $S(t)$. Some of the infected recover from the disease, and this confers immunity, while some die from the disease and are therefore no longer infectious. We count both of these as $R(t)$: the recovered, who are immune, and the dead, who can no longer transmit the disease. Write down the three coupled differential equations for $\frac{dI}{dt}$, $\frac{dS}{dt}$, and $\frac{dR}{dt}$.

There should only be two independent proportionality constants you can use in modeling this situation.

8. For the SIR model in exercise 7, draw the direction field with arrows indicating the direction of increase with time, in a plot of $I(t)$ vs. $S(t)$. Discuss what happens in time to the introduction of some infected individuals to a large population of susceptibles, and infer the condition under which an epidemic would develop.

9. *Two-sex models*

A process of modeling is to critically reexamine any model you have come up with, discover its deficiencies, and continually improve it.

It is obvious that the common Malthusian model on population growth is deficient:

$$\frac{dN}{dt} = BN - DN, \tag{9.36}$$

where $N(t)$ is the population density, B the per capita birthrate, and D the per capita death rate. It is deficient because it seems to imply that it is not only females that bear offspring.

A simple modification would have been:

$$\frac{dN}{dt} = Bf - DN, \tag{9.37}$$

in recognition of the fact that only the females can give birth. We can then develop a two-sex model for the male population ($m(t)$) and the female population ($f(t)$), which comprise the total population ($N(t) = m(t) + f(t)$). That model, however, also has a deficiency. The model (9.37) has the problem that there are births even if there are no males in the population! Obviously, (9.37) is appropriate only if the females are the limiting species (i.e., $f \ll m$). If the males are the limiting species ($m \ll f$), the births should be proportional to m:

$$\frac{dN}{dt} = Bm - DN. \tag{9.38}$$

It is desirable to have a model that works for any m, f densities and reduces to (9.37) if $f \ll m$ and to (9.38) if $m \ll f$. One such model is

$$\frac{dN}{dt} = B\frac{mf}{m+f} - DN. \tag{9.39}$$

This model is symmetric with respect to the two sexes.

If $m \ll f$, (9.39) reduces to (9.38). If $f \ll m$, (9.39) reduces to (9.37). Since there are two unknowns, we need to split (9.39) up into two equations:

$$\begin{aligned}
\frac{df}{dt} &= B_f\frac{mf}{m+f} - D_f f, \\
\frac{dm}{dt} &= B_m\frac{mf}{m+f} - D_m m,
\end{aligned} \tag{9.40}$$

where D_f is the female per capita death rate and D_m is the male per capita death rate; B_f is proportional to the female birthrate and B_m is proportional to the male birthrate. The birth ratio, $r \equiv \frac{\text{male births}}{\text{female births}}$, is given by

$$r = \frac{B_m\frac{mf}{m+f}}{B_f\frac{mf}{m+f}} = \frac{B_m}{B_f}.$$

This quantity is assumed constant in this model.

Use (9.40) to answer the following questions:

a. Try to find the nontrivial equilibrium male and female populations (m^*, f^*). It turns out that in general there is no nontrivial equilibrium unless certain conditions involving the parameters are met. Find that condition.

b. Even if m^* and f^* cannot reach equilibrium individually (if the condition in (a) is not satisfied), there is a well-defined equilibrium for the sex ratio. Calculate the equilibrium sex ratio $(m/f)^*$. Assume that the death rates, D_m and D_f, are the same.

c. For $D_f = D_m = D$, show that

$$\left[\frac{1}{B_f}f(t) - \frac{1}{B_m}m(t)\right] = \left[\frac{1}{B_f}f(0) - \frac{1}{B_m}m(0)\right]e^{-Dt}.$$

d. From (c) it is seen that after a few generations (with a generation defined by $D\Delta t \sim 1$), we practically have

$$m(t) \cong (B_m/B_f)\,f(t).$$

Substitute this into the female equation in (9.40) and solve for $f(t)$.

10. Another two-sex model that has been proposed is

$$\frac{df}{dt} = Bmf - Df,$$
$$\frac{dm}{dt} = Bmf - Dm.$$
(9.41)

This model, however, is problematic because it blows up in finite time.

a. Show, again, that the sex difference $(f(t) - m(t))$ becomes small in one or two generations.

b. So after a couple of generations, we can replace $m(t)$ by $f(t)$ approximately in the female equation in (9.41). This leads to

$$\frac{df}{dt} = -Df\left(1 - \frac{B}{D}f\right).$$
(9.42)

Show that for some t_0, if $f(t_0) < D/B$, the population will become extinct for t much larger than t_0.

c. Show that if $f(t_0) > D/B$, the solution will blow up in finite $(t - t_0)$. (Use separation of variables to solve (9.42).)

11. *Volterra equations*

Solve the Volterra equations in the form of (9.22):

$$\frac{dy}{dx} = \frac{y(bx - k)}{x(r - ay)},$$

subject to the condition that $y = y_0$ when $x = x_0$ (i.e., the trajectory at some time passes through the point (x_0, y_0)). Show that the solution (in the form of a relation between x, y, x_0, and y_0) implies that the trajectories $x = x(t)$, $y = y(t)$ are periodic if they do not pass through the equilibrium points. (*Hint*: Show that $y = y(t)$ is a simple closed curve.)

10

Marriage and Divorce

Mathematics required:
> same as the previous chapter

10.1 Introduction

About two thirds of marriages in the United States end in divorce within a 40-year period, with huge social and economic consequences. I cannot think of any form of legal contract, other than the marital vow, that sane adults would willingly enter into knowing full well such a high potential failure rate. The divorce rate for second marriages is even higher, about 75%; one would have thought that the participants would have become wiser from their first experience! Previous attempts at predicting marital dissolution tended to be based on mismatches in the couple's personality or modes of communication. These have not proved to be too successful.

Professor John Gottman of the University of Washington (Figure 10.1) does ground-breaking research on marriage, divorce, and repair. Gottman's work contradicts the "men are from Mars, women are from Venus" school of relationships, which holds that a lack of understanding of gender differences in communication styles is at the root of marital problems. His prediction of which couples would divorce within a four-year period is 94% accurate. Based on decades of experience interviewing couples, Gottman (1994) identified five types of married couples: Volatiles, Validators, Avoiders, Hostiles, and Hostile-detacheds. The first three produce stable marriages, while the last two produce unstable (high-risk) marriages. A Volatile couple tends to be quite romantic and passionate, but there are also heated arguments and fights. The Validating couple is calmer and intimate; these couples appear to place a high degree of value on shared experience, not on individuality. The Avoider couples avoid the pain of confrontation and conflict. The spouses interact with each other only in the positive range of their emotions. The Validating couple is the classic example of a successful marriage and is well known, but the surprise finding is that two other types of marriages are also stable. Another one of Gottman's unexpected

Figure 10.1. Professor John Gottman in his "Love Lab" at the University of Washington. (Courtesy of UW News and Information.)

findings is that anger is not the most destructive emotion in a marriage, since both happy and unhappy couples fight.

The types of marital interaction are deduced from interviews with the couples and subsequent observation (videotaped) and coding. The couple is asked to choose a problem area to discuss in a 15-minute session; details of the exchange are tracked by video cameras, and the emotions of both partners measured by electrodes taped to their foreheads and chests. The problem area could be the frequency of sex, money, in-laws, etc. The videotapes of the interactions as the couple works through the conflict are then coded. There are a number of different coding systems, but the objective is to measure positive versus negative responses in each spouse as he/she speaks in turn. Since all couples, even happily married ones, have some amount of negative interaction, and all couples, even unhappily married ones, have some degree of positive interaction, some averaging or smoothing of the scores is necessary. In the Rapid Couples Interaction Scoring System (RCISS) used by Gottman (1979), the total number of positive RCISS

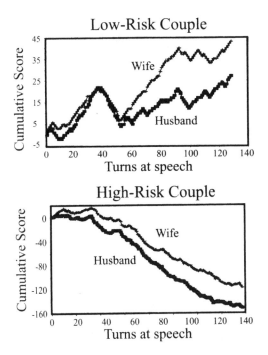

Figure 10.2. The cumulative RCISS scores of husband and wife for a typical low-risk couple (top panel) and for a high-risk couple (lower panel). (From Gottman et al. [2002].)

speaker codes (where the spouse says something positive) minus the total number of negative speaker codes was plotted for each spouse as a function of turns of speech (in essence, time). Two examples are given in Figure 10.2. The average slope of these cumulative scores is then used to classify the couples into high risk or low risk. Since the slope of the cumulated score is the same as the running mean of the score at each time, we shall now refer to (the slope of) the scores as measuring the (running time) average positivity or negativity of the spouse as a function of time.

The five types of couples seem to have different interaction styles. Figure 10.3 summarizes the different influence functions between husband and wife by fitting the empirical data into a two-slope functional form, as we will explain in a moment.

Validators have an influence function that tends to create a positive response in a spouse if the partner's behavior is positive and a negative response if the partner is negative. Volatiles and Conflict-avoiders have influence functions that appear to be one half of the Validators', with Volatiles having the right half of the curve close to the zero slope and the Avoiders having a near zero slope on the left half. Avoiders, by

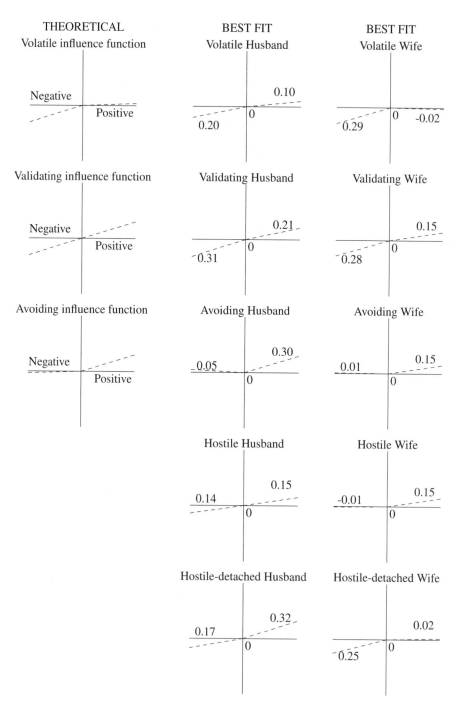

Figure 10.3. Influence functions. The left column is the theoretical influence function that summarizes the observation. The right two columns are best fits to the data using a two-slope linear function, with the numerical value of the slopes indicated. (Based on data in Cook et al. [1995].)

definition, tend not to interact with each other when they are negative; they influence one another only with positivity. On the other hand, the Volatile spouses influence one another primarily with negativity.

Hostile couples appear to have mixed a Validator husband influence function with an Avoider wife influence function. The Hostile-detached couples mix a Validator husband influence function with a Volatile wife influence function.

With three possible interaction styles for each of the two spouses in a marriage, there could in principle be nine, instead of five, types of marriages. Three of these have *matched* interaction styles (the same style for husband and wife) and six are *mismatched*. Four of the six mismatched marriages, e.g., one that mixes a Volatile style with an Avoider style, were not found in Gottman's data. Perhaps they are just too different for the relationship to survive, even temporarily, through courtship.

Of the five types of marriages observed, the stable marriages all seem to have matched styles of interaction between husband and wife, while the unstable marriages seem to have mismatched styles. Hence it was suggested (Cook et al., 1995) that perhaps the unstable marriages are simply failures to create a stable adaptation to a marriage that is matched, either Volatile, Validating, or Avoiding. For example, a person who is more suited to a Volatile or an Avoiding marriage may have married one who wishes for a Validating marriage, resulting in an unstable marriage.

Since there are many factors that can influence marital stability, it is not clear at this point that influence-style matches or mismatches are the determining factor. Another important factor appears to be an individual's own natural disposition, which measures how one reacts in the absence of spousal interaction. For Volatile couples, both husband and wife tend to have very positive personalities, followed by Validators and Avoiders. Volatile couples have stable marriages perhaps not because their influence styles are "matched" but because they are naturally very positive individuals; their marriage survives *despite* their influence style, which shows that they affect each other negatively. The husband and wife in the unstable marriages tend to have very negative natural dispositions, and *this* may be the reason for their marital problems.

10.2 Mathematical Modeling

We would like to mathematically model the interplay between the two factors affecting marriage and its dissolution: the natural disposition of husband and wife and the style of interaction each has. The aim of the mathematical model is to use the parameters distilled from the taped interviews to predict the long-term behavior of the marriage.

Let $x(t)$ be a measure of the husband's positivity (e.g., happiness) and $y(t)$ be the corresponding measure for the wife. In terms of the RCISS-coded scores, we are referring to its running mean and not the cumulative scores.

Self-Interaction

In the absence of marital interaction, e.g., when each was single, each spouse tends to his/her own "uninfluenced steady state": x_0 and y_0. This process is modeled by

$$\frac{dx}{dt} = r_1(x_0 - x),$$

$$\frac{dy}{dt} = r_2(y_0 - y).$$

The equilibrium (steady state) solution is

$$x^* = x_0, \quad y^* = y_0,$$

and perturbations from this equilibrium return to the equilibrium at an exponential decay rate of r_1 for the husband and r_2 for the wife. This aspect of the model can be seen by solving the above uncoupled ordinary differential equations in terms of

$$u = x - x_0, \quad v = y - y_0:$$

$$\frac{du}{dt} = -r_1 u, \quad \frac{dv}{dt} = -r_2 v,$$

giving

$$u(t) = u(0)e^{-r_1 t},$$

$$v(t) = v(0)e^{-r_2 t}.$$

The parameters for the uninfluenced states, x_0, y_0, r_1, and r_2, are determined from the subset of data points of the couple when the score of one of the spouses is zero. This happens in about 15% of the data (Cook et al., 1995). Presumably we can also determine these parameters if each spouse is interviewed alone.

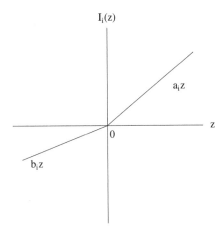

Figure 10.4. The theoretical influence function.

Marital Interactions

Cook et al. (1995) identified three major styles of marital interactions for husband, x, and for wife, y. We attribute the type of style of the influence function to the influencer. Here $I_2(x)$ is the influence exerted on the wife by the husband's emotions, while $I_1(y)$ is the influence of the wife on the husband (see Figure 10.4).

For all possible styles, the governing mathematical model is

$$\frac{dx}{dt} = r_1(x_0 - x) + I_1(y), \tag{10.1}$$

$$\frac{dy}{dt} = r_2(y_0 - y) + I_2(x), \tag{10.2}$$

where

$$I_i(z) = \begin{cases} a_i z & \text{if } z > 0, \\ b_i z & \text{if } z < 0. \end{cases}$$

Here i can be either 1 or 2, and z stands for either x or y.

In a *validating* style of interaction, one spouse influences the other across both the negative and positive ranges of emotions, with $a \cong b$; i.e., the slopes in the interaction function are approximately the same (see Figure 10.3). In a *conflict avoiding* style of interaction, the spouse who adopts this style avoids interacting with the other spouse through the negative range of his/her emotions, with $b/a \ll 1$. The spouse who

TABLE 10.1

Uninfluenced steady states and rates of relaxation for each spouse in five types of marriages, based on data from Cook et al. (1995)

Group	Husband's steady state			Wife's steady state		
	r_1	x_0	x^*	r_2	y_0	y^*
Low-risk couples						
Volatile	0.67	.68	.75	0.80	.68	.61
Validating	0.63	.38	.56	0.86	.52	.59
Avoiding	0.82	.26	.53	0.75	.46	.60
Average	0.71	.44	.61	0.80	.55	.60
High-risk couples						
Hostile	0.68	.10	.03	.51	−.64	−.45
Hostile-detached	0.60	−.42	−.50	.54	−.24	−.62
Average	0.64	−.16	−.24	.53	−.44	−.54

has a *volatile* style interacts with the other mainly in the negative range of his/her emotion, with $a/b \ll 1$.

10.3 Data

The parameters that enter into our model were determined empirically using codings of the interviews. They are converted from the discrete form used by Cook et al. (1995) into the form useful for our continuous model and listed in Table 10.1. x_0 and x^*, y_0 and y^* are in units of the mean RCISS scores, while r_1 and r_2 are in units of the RCISS scores per unit time (turns of speech).

It is seen that the uninfluenced steady states (i.e., x_0, y_0) appear to separate the high-risk couples from the low-risk couples. At least one of the spouses is highly negative in the high-risk couples. The wife is highly negative in the Hostile couple, while the husband is highly negative in the case of the Hostile-detached couple. On the other hand, the low-risk couples all have positive dispositions to various degrees. Both husband and wife in a Volatile couple have very positive natural dispositions, followed by the Validator couple, which is followed by the Avoider couple. It appears that the Volatile marriage is highly "regulated" by the couple's uninfluenced state, and the marriage is successful *despite* their negative marital interaction style. The spouses of a Validator couple are somewhat less positive than the Volatiles and they make up for it by interacting with each other positively as well as negatively. The Avoiders are less secure because they are even less positive than the Validators. They keep their marriages intact by avoiding conflict.

10.4 An Example of a Validating Couple

Before we proceed to solve the general case, let us consider the classic example of a *Validating* couple. It is known that their marriage is "regulated" (low risk). We want to know why, mathematically.

The couple has a *matched* style of interaction, with $a_i \sim b_i$. We shall in fact take $a_i = b_i$ for simplicity in this example. Thus

$$\frac{dx}{dt} = r_1(x_0 - x) + a_1 y, \tag{10.3}$$

$$\frac{dy}{dt} = r_2(y_0 - y) + a_2 x. \tag{10.4}$$

The husband is influenced by the wife's positivity (or negativity) with the rate of change due to this influence being a_1. Similarly, the wife is influenced by the husband's emotions, with a_2 being the constant of proportionality. Curiously, the set (10.3) and (10.4), governing the marital interactions, is of exactly the same form as the set of equations governing the *arms race* between two warring states (see exercise 1 of chapter 9)!

The long-term equilibrium state for a couple is found by setting the right-hand sides of (10.3) and (10.4) to zero. The equilibrium solution is denoted by (x^*, y^*):

$$r_1(x_0 - x^*) + a_1 y^* = 0, \quad r_2(y_0 - y^*) + a_2 x^* = 0.$$

The solution is easily found to be

$$x^* = \left[x_0 + \frac{a_1}{r_1} y_0\right] / \left[1 - \frac{a_1 a_2}{r_1 r_2}\right],$$

$$y^* = \left[y_0 + \frac{a_2}{r_2} x_0\right] / \left[1 - \frac{a_1 a_2}{r_1 r_2}\right]. \tag{10.5}$$

This single equilibrium turns out to be a *stable node* if

$$0 < (a_1 a_2)/(r_1 r_2) < 1.$$

Otherwise it is an *unstable saddle* (see chapter 9).

Generally, $r_1/a_1 > 1$, and $r_2/a_2 > 1$, since one responds to one's own feelings more quickly than to the spouse's feelings. Consequently, we are generally dealing with the stable case,

$$a_i/r_i < 1, \quad b_i/r_i < 1, \tag{10.6}$$

of a single equilibrium. (Incidentally, it is easy to understand why the case with $r_1/a_1 < 1$ and $r_2/a_2 < 1$ is unstable. The two spouses feed on each other's emotions without much of a moderating mechanism: the wife being sad makes the husband sadder, which in turn makes the wife even sadder, and so on. A similar chain of events happens when a spouse is happy. We will not consider this case in what follows. This unstable case, however, is applicable to *arms race* modeling.)

The long-term effects of marital interaction on the positivity (i.e., happiness) of each spouse in a Validating marriage are twofold: additive and magnifying. If both husband and wife are naturally positive people, i.e., $x_0 > 0$, $y_0 > 0$, their happiness is enhanced by the positivity of their spouse. This influence is additive, i.e.,

$$\text{Husband: Single:} \quad x_0 \rightarrow \text{married:} \quad x_0 + \frac{a_1}{r_1}y_0,$$

$$\text{Wife: Single:} \quad y_0 \rightarrow \text{married:} \quad y_0 + \frac{a_2}{r_2}x_0.$$

The sum, $x_0 + \frac{a_1}{r_1}y_0$, is also magnified by the factor $1/[1 - \frac{a_1 a_2}{r_1 r_2}] > 1$ in marriage. The result is that in marriage:

$$x^* = \left[x_0 + \frac{a_1}{r_1}y_0\right] \Big/ \left[1 - \frac{a_1 a_2}{r_1 r_2}\right] \gg x_0,$$

$$y^* = \left[y_0 + \frac{a_2}{r_2}x_0\right] \Big/ \left[1 - \frac{a_1 a_2}{r_1 r_2}\right] \gg y_0.$$

That is, the spouses are both much happier in marriage than if they were single. The Validating couple is said to have a *regulated* (low-risk) marriage. On the other hand, if the husband and wife both have negative "uninfluenced" steady state, i.e., $x_0 < 0$, $y_0 < 0$, the Validating marriage would make them more unhappy, i.e., $x^* < x_0$, $y^* < y_0$. This marriage is probably *unregulated* (high risk).

If x_0 and y_0 are of opposite sign, the marriage can still be successful under certain conditions. For example, if the husband is negative, i.e., $x_0 < 0$, but the wife is very positive, through marital interactions the husband's steady state, x^*, can be made positive if $x_0 + \frac{a_1}{r_1}y_0 > 0$. The wife's steady state, y^*, can still be positive if $y_0 + \frac{a_2}{r_2}x_0 > 0$. The marriage is probably successful.

Exact Solution

Since (10.3) and (10.4) are linear, they can in fact be solved exactly. The exact solution, with c_1 and c_2 being arbitrary constants determinable from initial conditions, is

$$x(t) = x^* + c_1 e^{\lambda_1 t} + c_2 e^{\lambda_2 t},$$

$$y(t) = y^* + c_1 \left(\frac{r_1}{a_1} + \frac{\lambda_1}{a_1} \right) e^{\lambda_1 t} + c_2 \left(\frac{r_1}{a_1} + \frac{\lambda_2}{a_1} \right) e^{\lambda_2 t},$$

where

$$\lambda_{1,2} = \frac{p}{2} \pm \frac{1}{2}\sqrt{p^2 - 4q}, \quad p \equiv -(r_1 + r_2), \quad q \equiv r_1 r_2 - a_1 a_2.$$

The solution is stable if $\lambda_1 < 0$, $\lambda_2 < 0$, and unstable if at least one of them is positive. Since p is negative, $\lambda_{1,2}$ are both negative if $q > 0$. That is the same as the stability criterion (10.6) mentioned previously. For the case of $q > 0$, the solution will eventually tend to x^*, y^*, the equilibrium solution given by (10.5).

10.5 Why Avoiding Conflicts Is an Effective Strategy in Marriage

When the spouses in a marriage do not have as positive a natural disposition as the Validating couple, adopting a conflict-avoiding style of interaction may be effective in maintaining a successful marriage. This is a surprising finding since traditionally psychologists have thought that sweeping marital problems under the rug, where they fester, may be detrimental to the health of a marriage. We model an Avoider couple by the following model:

$$\frac{dx}{dt} = r_1(x_0 - x) + a_1 I(y),$$

$$\frac{dy}{dt} = r_2(y_0 - y) + a_2 I(x),$$

where $I(y)$ is defined as $I(y) = y$ if $y > 0$, $I(y) = 0$ if $y < 0$, and similarly for $I(x)$.

To find the equilibrium solution, we set the right-hand sides of the above two equations to zero, i.e.,

$$r_1(x_0 - x^*) + a_1 I(y^*) = 0,$$

$$r_2(y_0 - y^*) + a_2 I(x^*) = 0.$$

If $x^* > 0$, $y^* > 0$, then $I(x^*) = x^*$, $I(y^*) = y^*$, and so the equilibrium solution is the same as that for the Validating couple:

$$x^* = \left[x_0 + \frac{a_1}{r_1} y_0 \right] \bigg/ \left[1 - \frac{a_1 a_2}{r_1 r_2} \right] > 0,$$

$$y^* = \left[y_0 + \frac{a_2}{r_2} x_0 \right] \bigg/ \left[1 - \frac{a_1 a_2}{r_1 r_2} \right] > 0.$$

On the other hand, if $x^* < 0$, $y^* < 0$, then $I(x^*) = 0$, $I(y^*) = 0$, and we have

$$x^* = x_0 < 0, \quad y^* = y_0 < 0.$$

It can be seen that if the couple are individually negative (i.e., $x_0 < 0$, $y_0 < 0$), they are no less unhappy in marriage ($x^* = x_0$, $y^* = y_0$). However, if they are naturally positive (i.e., $x_0 > 0$, $y_0 > 0$), they are much happier in marriage ($x^* > x_0 > 0$, $y^* > y_0 > 0$). Thus a conflict-avoiding style is a "safer" style of marital interaction for the couples whose natural dispositions are not as high as those of Validating couples.

The conflict-avoiding marriage is also a successful one even if one spouse is negative, provided that the other spouse is more positive than the negative spouse is negative. By avoiding interaction when one is negative, the couple feeds on the positive spouse's positivity and is not affected by the negative spouse's negativity. The steady state is the same as that for the Validating couple if at least one spouse is positive (if $x_0 + \frac{a_1}{r_1} y_0 > 0$, $y_0 + \frac{a_2}{r_2} x_0 > 0$).

10.6 Terminology

This is probably a good time to define what we mean by a "successful" or "failed" marriage, and by a "stable" or "unstable" marriage. The words "stable" and "unstable" have different meanings to a mathematician and to a behavioral psychologist. We say a marriage is successful ("regulated") if the couple is happier in marriage than if single. We may want to relax this condition a little, to say that each spouse is at least as happy in marriage as they would be if single. Thus, the condition for a successful marriage is:

$$x^* \gtrsim x_0, \quad y^* \gtrsim y_0.$$

As we have found, the equilibrium (x^*, y^*) is stable. The solution will inevitably tend to this equilibrium, unless actions not described by our model intervene. If the marriage is tending to a steady state with $x^* < 0$, $y^* < 0$ (both husband and wife unhappy), one or both spouses may wake up one day and make a decision that he/she does not want this marriage to continue this way any longer. This action of dissolving the marriage is not contained in our model. We have no reason to expect that this action will manifest itself as an "instability" of our solution. Instead, we simply predict, based on the fact that the long-term state to which the marriage is tending is so negative, that the probability is high that one or both of the spouses will take the action to dissolve the marriage.

10.7 General Equilibrium Solutions

We now solve (10.1) and (10.2) for the equilibrium solution obtained by setting ($\frac{dx}{dt} = 0$, $\frac{dy}{dt} = 0$). This equilibrium is given by

$$
\begin{aligned}
r_1(x_0 - x^*) + I_1(y^*) = 0, \\
r_2(y_0 - y^*) + I_2(x^*) = 0.
\end{aligned}
\tag{10.7}
$$

There is only a single equilibrium solution (x^*, y^*) of (10.7). When (10.6) is satisfied, this equilibrium is stable. This stable equilibrium can be found in any one of the four quadrants depending on the parameters, x_0, y_0, r_1, r_2, and a_i and b_i. The successful marriages are generally found in the first quadrant, where $x^* > 0$, $y^* > 0$; failed marriages are generally located in the third quadrant, where $x^* < 0$, $y^* < 0$.

1. *First quadrant, successful marriages*
   For $x^* > 0$, $y^* > 0$, (10.7) becomes

$$
x^* = x_0 + \frac{1}{r_1} I_1(y^*) = x_0 + \frac{1}{r_1} a_1 y^*,
$$

$$
y^* = y_0 + \frac{1}{r_2} I_2(x^*) = y_0 + \frac{1}{r_2} a_2 x^*.
$$

So, solving them simultaneously yields

$$
\begin{aligned}
x^* = \left[x_0 + \frac{a_1}{r_1} y_0 \right] \Big/ \left[1 - \frac{a_1 a_2}{r_1 r_2} \right] > 0, \\
y^* = \left[y_0 + \frac{a_2}{r_2} x_0 \right] \Big/ \left[1 - \frac{a_1 a_2}{r_1 r_2} \right] > 0,
\end{aligned}
\tag{10.8}
$$

In order for x^*, y^* to both be positive, we need

$$x_0 + \frac{a_1}{r_1} y_0 > 0, \quad y_0 + \frac{a_2}{r_2} x_0 > 0 \tag{10.9}$$

(assuming that (10.6) is satisfied).

For the case where (10.9) is true, we have

$$x^* \gg x_0 > 0, \quad y^* \gg y_0 > 0.$$

Both the additive and magnifying effects of marital interaction are present. The marriage is a successful one.

One surprising conclusion from the above result is that the marriage is successful *no matter what interaction style each spouse adopts*, as long as each's natural disposition is positive ($x_0 > 0$, $y_0 > 0$). This is consistent with the data in Table 10.1.

2. *Third quadrant, unsuccessful marriages*

The equilibrium solution is

$$
\boxed{
\begin{aligned}
x^* &= \left[x_0 + \frac{b_1}{r_1} y_0 \right] \Big/ \left[1 - \frac{b_1 b_2}{r_1 r_2} \right] < 0, \\[2mm]
y^* &= \left[y_0 + \frac{b_2}{r_2} x_0 \right] \Big/ \left[1 - \frac{b_1 b_2}{r_1 r_2} \right] < 0,
\end{aligned}
}
\tag{10.10}
$$

provided that

$$x_0 + \frac{b_1}{r_1} y_0 < 0, \quad y_0 + \frac{b_2}{r_2} x_0 < 0. \tag{10.11}$$

One conclusion is that if $x_0 < 0$, $y_0 < 0$, the marriage is doomed to failure no matter what interaction style each spouse adopts. Any marital interaction will lead to a more negative x^*, y^*. The Hostile marriage is characterized by the wife's excessive natural negativity ($y_0 < 0$). She tries to avoid conflict by interacting only in the positive range of her emotions, but she does not have very much of that. The marriage is doomed to failure regardless of the husband's interaction style. In the Hostile-detached couples, the marriage fails because the husband is so naturally negative ($x_0 < 0$). The matter is made worse by the failure of both spouses to avoid conflict.

3. *Second quadrant*

$$x^* < 0, \quad y^* > 0.$$

The equilibrium solution is

$$
x^* = \left[x_0 + \frac{a_1}{r_1} y_0 \right] \Big/ \left[1 - \frac{a_1 b_2}{r_1 r_2} \right] < 0,
$$
$$
y^* = \left[y_0 + \frac{b_2}{r_2} x_0 \right] \Big/ \left[1 - \frac{a_1 b_2}{r_1 r_2} \right] > 0,
$$

(10.12)

provided that

$$
x_0 + \frac{a_1}{r_1} y_0 < 0, \quad y_0 + \frac{b_2}{r_2} x_0 > 0.
$$

Even though the husband's steady state is negative, we would still classify the marriage successful if

$$
x^* > x_0, \quad y^* \cong y_0 > 0.
$$

For a naturally negative husband ($x_0 \leq 0$) and a positive wife ($y_0 > 0$) we always have $x^* > x_0$ regardless of the style of interaction the wife adopts. However, in order that the wife not be too influenced by the husband's negativity, he should adopt a conflict-avoider style of interaction ($b_2 \cong 0$).

If, on the other hand, the wife is naturally negative ($y_0 < 0$) and the husband positive ($x_0 > 0$), the marriage can still be successful if a_1/r_1 is small but b_2/r_2 large. In that case, we would have

$$
x^* \cong x_0 > 0, \quad y^* > y_0.
$$

4. *Fourth quadrant*

$$
x^* > 0, \quad y^* < 0.
$$

This case is similar to case (3), except with the roles of the husband and wife reversed. That is,

$$
x^* = \left[x_0 + \frac{b_1}{r_1} y_0 \right] \Big/ \left[1 - \frac{b_1 a_2}{r_1 r_2} \right] > 0,
$$

$$
y^* = \left[y_0 + \frac{a_2}{r_2} x_0 \right] \Big/ \left[1 - \frac{b_1 a_2}{r_1 r_2} \right] < 0,
$$

provided that $x_0 + \frac{b_1}{r_1} y_0 > 0$ and $y_0 + \frac{a_2}{r_2} x_0 < 0$.

10.8 Conclusion

From taped interviews with couples, we attempt to deduce the parameters that we need for our mathematical model, such as the style of interaction of each spouse (a_i, b_i), their uninfluenced steady state (x_0, y_0), and each's inertia to change $(1 - r_i)$. Since it is not feasible to observe the couple for long periods of time, we rely on the mathematical model to predict where the couple will tend if the marriage is allowed to evolve with these parameters. If the long-term solution of the model is such that the husband and wife are tending to the negative quadrant $(x^* < 0, y^* < 0)$, we then predict that the current marriage is heading for dissolution if the parameters are not changed. The mathematical stability of the equilibrium is not predictive of marital stability since all equilibria in this model are stable.

The future direction of research focuses on marriage repair and, related to it, an understanding of why a couple develops a particular interaction style in marriage. These issues may need to be addressed using the mathematics of *game theory*.

10.9 Assignment

To predict the long-term outcome of a particular marriage, we need to first observe the couple and determine empirically the parameters that go into our mathematical model. For this assignment, we have five couples whose parameters have been determined. They happen to be the same as those listed in Table 10.1 and Figure 10.3 (which were actually for the mean of five types of couples).

Let $x(t)$ be the husband's RCISS score and $y(t)$ be the wife's RCISS score. RCISS scores are a measure of a spouse's positivity in marriage. The equations governing $x(t)$ and $y(t)$ are

$$\frac{dx}{dt} = r_1(x_0 - x) + I_1(y),$$

$$\frac{dy}{dt} = r_2(y_0 - y) + I_2(x),$$

where

$$I_1(z) = \begin{cases} a_i z & \text{if } z > 0, \\ b_i z & \text{if } z < 0. \end{cases}$$

The slopes a_i and b_i for various couples are given in Figure 10.3. The parameters $r_1, r_2, x_0,$ and y_0 are given in Table 10.1.

Determine the long-term values of $x(t)$ and $y(t)$ (i.e., x^* and y^*) for each of the five types of couples: Volatile, Validating, Avoider, Hostile, and Hostile-detached. (First solve x^*, y^* in algebraic form; then put in the numbers to get numerical values for each case. Do not use the values of x^*, y^* in Table 10.1.)

Solution
Find the equilibrium x^*, y^* from

$$0 = r_1(x_0 - x^*) + I_1(y^*),$$
$$0 = r_2(y_0 - y^*) + I_2(x^*).$$

From the "general equilibrium solution" section of this chapter, we have the following.

I. If $x_0 + \frac{a_1}{r_1}y_0 > 0$, $y_0 + \frac{a_2}{r_2}x_0 > 0$, then

$$x^* = \left[x_0 + \frac{a_1}{r_1}y_0\right] \Big/ \left[1 - \frac{a_1 a_2}{r_1 r_2}\right],$$

$$y^* = \left[y_0 + \frac{a_2}{r_2}x_0\right] \Big/ \left[1 - \frac{a_1 a_2}{r_1 r_2}\right].$$

II. If $x_0 + \frac{a_1}{r_1}y_0 < 0$, $y_0 + \frac{b_2}{r_2}x_0 > 0$, then

$$x^* = \left[x_0 + \frac{a_1}{r_1}y_0\right] \Big/ \left[1 - \frac{a_1 b_2}{r_1 r_2}\right],$$

$$y^* = \left[y_0 + \frac{b_2}{r_2}x_0\right] \Big/ \left[1 - \frac{a_1 b_2}{r_1 r_2}\right].$$

III. If $x_0 + \frac{b_1}{r_1}y_0 < 0$, $y_0 + \frac{b_2}{r_2}x_0 < 0$, then

$$x^* = \left[x_0 + \frac{b_1}{r_1}y_0\right] \Big/ \left[1 - \frac{b_1 b_2}{r_1 r_2}\right],$$

$$y^* = \left[y_0 + \frac{b_2}{r_2}x_0\right] \Big/ \left[1 - \frac{b_1 b_2}{r_1 r_2}\right].$$

IV. If $x_0 + \frac{b_1}{r_1} y_0 > 0$, $y_0 + \frac{a_2}{r_2} x_0 < 0$, then

$$x^* = \left[x_0 + \frac{b_1}{r_1} y_0 \right] \bigg/ \left[1 - \frac{b_1 a_2}{r_1 r_2} \right],$$

$$y^* = \left[y_0 + \frac{a_2}{r_2} x_0 \right] \bigg/ \left[1 - \frac{b_1 a_2}{r_1 r_2} \right].$$

All equilibria are stable.

From the data given, we have the following information about the various types of couples.

Volatile couple:

$$x_0 = 0.68, \quad y_0 = 0.68,$$
$$r_1 = 0.67, \quad r_2 = 0.80,$$
$$a_1 = -0.02, \quad a_2 = 0.10,$$

$$x_0 + \frac{a_1}{r_1} y_0 = 0.68 + \frac{(-0.02)}{0.67} \cdot 0.68 = 0.66 > 0,$$

$$y_0 + \frac{a_2}{r_2} x_0 = 0.68 + \frac{0.10}{0.80} \cdot 0.68 = 0.77 > 0.$$

This belongs to case (I) and so

$$x^* = 0.68/1.00 = 0.68,$$
$$y^* = 0.77/1.00 = 0.77.$$

This is a successful marriage.

Validating couple:

$$x_0 = 0.38, \quad y_0 = 0.52,$$
$$r_1 = 0.63, \quad r_2 = 0.86,$$
$$a_1 = 0.15, \quad a_2 = 0.21,$$

$$x_0 + \frac{a_1}{r_1} x_0 = 0.38 + \frac{0.15}{0.63} \cdot 0.52 = 0.50 > 0,$$

$$y_0 + \frac{a_2}{r_2} x_0 = 0.52 + \frac{0.21}{0.86} \cdot 0.38 = 0.61 > 0.$$

This belongs to case (I). $[1 - \frac{a_1 a_2}{r_1 r_2}] = 0.94$.

$$x^* = 0.50/0.94 = 0.53, \quad y^* = 0.61/0.94 = 0.65.$$

This is a successful marriage.

Avoiding couple:

$$x_0 = 0.26, \quad y_0 = 0.46,$$
$$r_1 = 0.82, \quad r_2 = 0.75,$$
$$a_1 = 0.15, \quad a_2 = 0.30,$$

$$x_0 + \frac{a_1}{r_1} y_0 = 0.26 + \frac{0.15}{0.81} \cdot 0.46 = 0.34 > 0,$$

$$y_0 + \frac{a_2}{r_2} x_0 = 0.46 + \frac{0.30}{0.75} \cdot 0.26 = 0.56 > 0.$$

This belongs to case (I). $[1 - \frac{a_1 a_2}{r_1 r_2}] = [1 - \frac{0.15 \times 0.30}{0.82 \times 0.75}] = 0.93$.

$$x^* = 0.34/0.93 = 0.37, \quad y^* = 0.56/0.93 = 0.60.$$

Hostile couple:

$$x_0 = 0.10, \quad y_0 = -0.64,$$
$$r_1 = 0.68, \quad r_2 = 0.51,$$
$$a_1 = 0.15, \quad a_2 = 0.15,$$
$$b_1 = -0.01, \quad b_2 = 0.14,$$

$$x_0 + \frac{b_1}{r_1} y_0 = 0.10 + \frac{(-0.01)}{0.68}(-0.64) = 0.11 > 0,$$

$$y_0 + \frac{a_2}{r_2} x_0 = -0.64 + \frac{0.15}{0.51} \cdot 0.10 = -0.61 < 0.$$

This belongs to case (IV).

$$x^* = \left[x_0 + \frac{b_1}{r_1} y_0 \right] \Big/ \left[1 - \frac{b_1 a_2}{r_1 r_2} \right] = 0.11/1.14 = 0.1,$$

$$y^* = \left[y_0 + \frac{a_2}{r_2} x_0 \right] \Big/ \left[1 - \frac{b_1 a_2}{r_1 r_2} \right] = -0.61/1.14 = -0.54.$$

This is a high-risk couple, but they are no more unhappy in marriage than when they are each alone.

Hostile-detached:

$$x_0 = -0.42, \quad y_0 = -0.24,$$

$$r_1 = 0.60, \quad r_2 = 0.54,$$

$$b_1 = 0.25, \quad b_2 = 0.17,$$

$$x_0 + \frac{b_1}{r_1} y_0 = -0.42 + \frac{0.25}{0.60} \cdot (-0.24) = -0.52 < 0,$$

$$y_0 + \frac{b_2}{r_2} x_0 = -0.24 + \frac{0.17}{0.54} \cdot (-0.42) = -0.37 < 0.$$

This belongs to case (III).

$$x^* = \left[x_0 + \frac{b_1}{r_1} y_0 \right] \Big/ \left[1 - \frac{b_1 b_2}{r_1 r_2} \right]$$

$$= -0.52/0.87 = -0.60,$$

$$y^* = \left[y_0 + \frac{b_2}{r_2} x_0 \right] \Big/ \left[1 - \frac{b_1 b_2}{r_1 r_2} \right]$$

$$= -0.37/0.87 = -0.43.$$

This is a high-risk couple; both husband and wife are very unhappy in marriage, more unhappy than when they are alone.

10.10 Exercises

1.   Let $x(t)$ and $y(t)$ be measures of happiness for the husband and wife, respectively. Negative values indicate unhappiness. Let x_0 and y_0 be the "natural disposition" of the husband and wife, respectively. This is how happy they would be if they were single. During marriage, the couple develops a style of interaction that is called "validating." A model of their marriage dynamics is:

$$\frac{dx}{dt} = r_1(x_0 - x) + a_1 y,$$

$$\frac{dy}{dt} = r_2(y_0 - y) + a_2 x,$$

where a_1 measures how easily the husband is influenced by the wife's emotions, and a_2 is the corresponding quantity for the wife:

$$0 < a_1/r_1 < 1, \quad 0 < a_2/r_2 < 1.$$

a. Find out where this marriage is heading.

b. A marriage is termed "regulated" and is low risk if the long-term happiness of each spouse is enhanced by the marital interaction. Otherwise it is called "unregulated" (high risk). Give reasons for your answers to the following questions:
   i. Is the marriage "regulated" if each of the spouses is naturally happy ($x_0 > 0$, $y_0 > 0$)?
   ii. What if $x_0 < 0$, $y_0 < 0$?

2.   The equations governing the marital interaction for a Validating couple, Eqs. (10.3) and (10.4), are exactly the same as those governing Richardson's model for the arms race between two warring states (exercise 1 of chapter 9). Discuss why in one case the married couple lives happily ever after and in the other case the arms race escalates into war. Use reasonable parameter values in your discussion.

3.   A Hostile-detached couple has a husband who is naturally very negative ($x_0 \ll 0$) and a wife who is also negative ($y_0 < 0$). The husband interacts with a validating style while the wife interacts with a volatile style:

$$\frac{dx}{dt} = r_1(x_0 - x) + a_1 J(y),$$

$$\frac{dy}{dt} = r_2(y_0 - y) + a_2 x,$$

where $x(t)$ denotes the husband's happiness and $y(t)$ the wife's happiness. $J(y) = 0$ if $y > 0$ but $J(y) = y$ if $y < 0.$ $0 < a_1/r_1 < 1, 0 < a_2/r_2 < 1$.

a. Show that if $x_0 < 0$, $y_0 < 0$, the couple is even more unhappy in marriage (at equilibrium).

b. Show that their style of integration can lead to a successful marriage if they are naturally positive (i.e., $x_0 > 0$, $y_0 > 0$). Compare your solution for this case to that of a Validating couple worked out in class. If the solutions are the same, explain why they don't differ. If the solutions differ, also explain why.

11

Chaos in Deterministic Continuous Systems, Poincaré and Lorenz

Mathematics introduced:
system of three coupled ordinary differential equations; linear and nonlinear stability; periodic and aperiodic solutions; deterministic chaos; sensitivity to initial conditions

11.1 Introduction

Unlike quantum mechanics, the world of Newtonian mechanics is described by deterministic equations: given a precise initial condition of a particle's position and velocity, these equations predict the precise trajectory of that particle for all later times. This does not mean, however, that a small difference in the initial condition will not lead to wildly different later trajectories.

These "chaotic," unpredictable behaviors were accidentally discovered by Henri Poincaré and Edward Lorenz in different contexts.

11.2 Henri Poincaré

For two hundred years after Isaac Newton (1642–1727) discovered his law of gravitation and law of motion (mass times acceleration is equal to force), it was not recognized that planetary motion governed by these two deterministic laws may yield "unpredictable" trajectories. The two-body problem (e.g., a planet revolving around a sun) was solved by Newton analytically. It yields an elliptic orbit for the planet, as observed by Johannes Kepler (1571–1630). The three-body problem, involving, e.g., a sun, a planet, and a moon, is much more difficult. We now know that this problem cannot be solved exactly (in closed form).

In 1885, to celebrate the 60th birthday of Oscar II, King of Sweden and Norway, a contest was proposed by the mathematician Magnus Gösta Mittag-Liffler (1846–1927). The problems, four of them, were proposed by his teacher Karl Weierstrass at the University of Berlin. One of the problems was on celestial mechanics and the stability

Figure 11.1. Henri Poincaré (1647–1727).

of the solar system:

> Given a system of arbitrarily many mass points that attract each other according to Newton's laws, assuming that no two points ever collide, give the coordinates of the individual points for all time as the sum of a uniformly convergent series whose terms are made up of known functions.

The contest attracted an entry from a brilliant young professor, Henri Poincaré (1854–1912) of the University of Paris (Figure 11.1).

Poincaré was probably the last universalist in mathematics, equally at home in pure and applied mathematics. His philosophy towards mathematics, however, is more akin to that of an applied mathematician, in that he believed in the important role of intuition in mathematics rather than in treating mathematics as deducible from logic. Poincaré had poineered in his doctoral thesis the geometric view for understanding differential equations qualitatively using phase spaces and trajectories, which is still being used to analyze systems of ordinary equations (called dynamical systems). He was anxious to try out his novel approach to celestial mechanics and submitted a lengthy entry on the three-body problem. Although it did not provide a complete solution to the question posed, it was deemed by Weierstrass as "nevertheless of such importance that its publication will inaugurate a new era in the history of celestial mechanics." Poincaré won the gold medal prize.

Figure 11.2. Professor Edward Lorenz. (Courtesy of E. N. Lorenz.)

While his paper was being readied for publication in *Acta Mathematica*, Poincaré discovered that the result on the stability of the three-body problem was in error. The press was stopped and copies already distributed were retrieved and destroyed, while Poincaré frantically worked on a massive revision of his submitted work. He eventually came to accept the possibility that the deterministic world of Newtonian mechanics admits the possibility of unpredictable solutions even for a system as simple as three gravitating bodies. In one geometric case he had not considered previously, Poincaré now found orbits "so tangled that I cannot even begin to draw them." Moreover, he discovered that "It may happen that small differences in the initial positions may lead to enormous differences in the final phenomena. Prediction becomes impossible." We now recognize this as the hallmark of a chaotic dynamical system, although the quoted sentence by Poincaré was a philosophical remark. Poincaré's mathematical discovery was largely neglected outside the celestial mechanics area until Edward Lorenz demonstrated the same ultrasensitivity to initial conditions in a simpler set of three ordinary differential equations.

11.3 Edward Lorenz

Edward Lorenz (Figure 11.2), a professor of meteorology at MIT, published his seminal paper on chaos in 1963 in the *Journal of Atmospheric Sciences*. That paper's impact on mathematics extends well beyond its original focus on weather prediction.

Lorenz discovered this "chaotic" behavior in 1959 (before the mathematical use of the word chaos was coined) when he was numerically integrating a truncated model of weather. Lorenz has the habit of doing all the programming and analyzing rows of numerical output personally, unlike most other professors then and now, who delegate such "routine" chores to postdocs and graduate students. He actually had one desk-size "personal" computer installed in this office for his own use. On this one occasion he wanted to examine more closely a particular segment of the printed output of a previous run. He thought he could restart the run in the middle by typing into the computer the model output as the initial condition. To his surprise he found that the restarted run diverged from the original run. He traced this discrepancy to the fact that while the computer had six digits of accuracy, the printed output had only three. That discrepancy in the last few digits of the solution led to major differences in the future behavior of the solution. Later Lorenz used this sensitivity to initial data as the basis for his argument that weather cannot be predicted in detail for more than a couple of weeks in advance. This is because, among other inaccuracies in our knowledge, we cannot measure wind and temperature initial conditions to very high degrees of accuracy. Lorenz found the same behavior later in a simplified three-component system.

Chaotic solutions are actually quite rare in differential equations. Their existence requires at least three degrees of freedom (three coupled first-order ordinary differential equations or a single second-order ordinary differential equation with an imposed forcing frequency) and nonlinearity. The simplest such model is the three-component truncated model of the Rayleigh–Bénard convection of Lorenz (1963). Even in such models, chaotic solutions exist under a very restricted set of model parameters. Such models are not solvable analytically. This may account for why chaotic solutions remained hidden for so long in the deterministic solutions to Newton's equation.

In 1990, Lorenz was invited to give a series of three Danz lectures at the University of Washington. (I had the distinct honor of introducing him to an audience of 500 in Kane Hall.) The lecture notes were later, in 1993, published in book form under the title *The Essence of Chaos*. In it Lorenz described how he arrived at the set of parameter values that give rise to chaotic solutions. Unlike Poincaré, who was not looking for chaos in realistic three-body systems such as the sun, earth, and moon and who had to reluctantly admit that a chaotic solution is possible under some unusual arrangement of parameters, Lorenz was at the time actively looking for model parameters that could produce fluctuations like those of temperature or pressure in real weather, which are neither steady nor periodic. He was using his model equations not to reproduce real weather but to produce a time series (a sequence of numbers) to test some statistical schemes of weather prediction. If his time series converged to a fixed point (an equilibrium) it would not be of any use to him because predicting the subsequent evolution of his "weather" would be unrealistically easy. Lorenz would change the constants in his equations to get more interesting "data." A periodic oscillation also was not satisfactory. It took many adjustments to the parameters for the time series to look interesting to Lorenz, i.e., aperiodic. Whether the parameters were realistic was not a concern to Lorenz.

11.4 The Lorenz Equations

As mentioned above, for (continuous) first-order ordinary differential equations, chaos normally does not exist in the solution unless there are at least three such equations nonlinearly coupled together. In the late 1950s, Professor Edward Lorenz was looking for a simple system of equations whose solution may possess aperiodic behavior. He was interested in having such a simple set to generate artificial data to test some statistical methods used in numerical weather prediction. A colleague showed Lorenz a seventh-order system (a system involving seven first-order ordinary differential equations) that was a spectrally truncated model for thermal convection in a fluid confined between two flat plates, the lower plate being maintained at a higher temperature than the upper one. The numerical solution of that seven-component system appeared to possess aperiodic behavior, which Lorenz called "deterministic nonperiodic flow." Since four of the variables eventually approach zero, Lorenz set these to zero and arrived at the following simpler three-component system, the now famous *Lorenz equations*:

$$
\begin{aligned}
\frac{dx}{dt} &= \sigma(-x + y), \\
\frac{dy}{dt} &= rx - y - xz, \\
\frac{dz}{dt} &= -bz + xy.
\end{aligned}
\tag{11.1}
$$

The variable $x(t)$ measures the rotational speed of the convection cell, with its sign indicating clockwise or anticlockwise rotation. The horizontal temperature variation is projected onto a single sinusoidal spatial

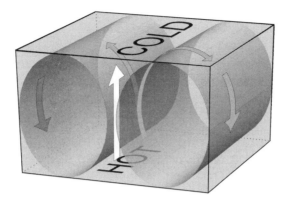

Figure 11.3. Two-dimensional convection cell.
(Drawing by Wm. Dickerson.)

pattern, with $y(t)$ measuring the difference in the temperature between the rising branch and the sinking branch of the convection cell. See Figure 11.3. In the absence of convection, the static heat conduction state has $x(t) \equiv 0$ and $y(t) \equiv 0$, with the vertical temperature being a linear function joining the temperature on the bottom plate to that on the upper plate. The variable $z(t)$ measures the deviation of the temperature from the linear conduction vertical profile, and $z(t) \equiv 0$ when there is no convection. The parameters in the equations are all positive and real. The Prandtl number σ describes the property of the fluid and the parameter b describes the geometry. The Rayleigh number r measures the specified temperature difference between the lower and upper plates. It is normalized so that $r = 1$ corresponds to the critical value for convection to first occur.

By heating the lower plate, r increases, and after exceeding a certain critical value ($r = 1$), convection of the fluid is needed to transport the heat from the lower plate to the upper plate ($x \not\equiv 0$). Before that, heat is simply conducted by a static fluid ($x \equiv 0$).

These results can be inferred from an examination of the equilibrium points of the Lorenz equations. Setting the right-hand side of (11.1) to zero, we have

$$\sigma(x^* - y^*) = 0,$$

$$rx^* - y^* - x^*z^* = 0,$$

$$-bz^* + x^*y^* = 0.$$

From the first equation we have $y^* = x^*$. Substituting it into the second and third equations, we get

$$x^*(r - 1 - z^*) = 0,$$

$$-bz^* + x^{*2} = 0.$$

Thus either $x^* = 0$ or $z^* = r - 1$. If the former, then $y^* = x^* = 0$ and $bz^* = x^{*2} = 0$, resulting in one equilibrium:

$$P_1 = (0, 0, 0).$$

If the latter, then $x^* = \pm\sqrt{bz^*} = \pm\sqrt{b(r - 1)} = y^*$, giving two more equilibria:

$$P_2 = (\sqrt{b(r - 1)}, \sqrt{b(r - 1)}, r - 1) \text{ and}$$

$$P_3 = (-\sqrt{b(r - 1)}, -\sqrt{b(r - 1)}, r - 1).$$

Since we are only interested in real equilibrium points, there is only one such equilibrium when $r < 1$:

$$P_1 = (0, 0, 0),$$

and this equilibrium is stable (see later). This is the static conduction state. When $r > 1$, P_1 becomes unstable and convection begins. It evolves into one or the other stable equilibrium pattern, P_2 or P_3. These eventually also become unstable for $r > r_c$. This is discussed further below.

To find the stability of an equilibrium (x^*, y^*, z^*), we perturb around it by assuming

$$(x(t), y(t), z(t)) = (x^* + u(t), y^* + v(t), z^* + w(t)),$$

with

$$\mathbf{u}(t) \equiv (u(t), v(t), w(t))$$

being the small perturbation.

The Lorenz system (11.1) becomes, upon dropping the products of u, v, w, which are small,

$$\frac{d}{dt}\mathbf{u}(t) = A\mathbf{u}(t), \tag{11.2}$$

where the constant matrix A is given by

$$A = \begin{bmatrix} -\sigma & \sigma & 0 \\ r - z^* & -1 & -x^* \\ y^* & x^* & -b \end{bmatrix}.$$

This system of linear ordinary differential equations with constant coefficients possesses solutions of the form

$$\mathbf{u}(t) = \mathbf{u}(0)e^{\lambda t}.$$

Substituting into Eq. (11.2) yields

$$\lambda \mathbf{u}(0) = A\mathbf{u}(0),$$

which is

$$(A - \lambda I)\mathbf{u}(0) = 0,$$

where I is a 3×3 unit matrix. The above algebraic system has the trivial solution

$$\mathbf{u}(t) = \mathbf{u}(0) \equiv 0.$$

Nontrivial solutions are possible only if

$$\det(\mathbf{A} - \lambda I) = 0.$$

That is,

$$\det \begin{bmatrix} -\sigma - \lambda & \sigma & 0 \\ r - z^* & -1 - \lambda & -x^* \\ y^* & x^* & -b - \lambda \end{bmatrix} = 0. \tag{11.3}$$

$(x^*, y^*, z^*) = (0, 0, 0)$
Equation (11.3) can be written as

$$(\lambda + b)(\lambda + \sigma)(\lambda + 1) - (\lambda + b)\sigma r = 0.$$

Factoring out $(\lambda + b)$:

$$(\lambda + b)[(\lambda + \sigma)(\lambda + 1) - \sigma r] = 0.$$

The three *eigenvalues* are

$$\lambda = \lambda_1 = -b,$$

$$\lambda = \lambda_2 = \frac{1}{2}[-(\sigma + 1) + \sqrt{(\sigma + 1)^2 - 4\sigma(1 - r)}],$$

$$\lambda = \lambda_3 = \frac{1}{2}[-(\sigma + 1) - \sqrt{(\sigma + 1)^2 - 4\sigma(1 - r)}].$$

The last two are the roots of the quadratic equation inside the square brackets. For $r < 1$, both λ_2 and λ_3 are real and negative. Since λ_1 is real and negative, the equilibrium point $P_1 = (0, 0, 0)$ is stable. For $r > 1$, λ_2 becomes positive, while λ_1 and λ_3 remain negative. *Therefore P_1 becomes unstable for $r > 1$.*

$(x^*, y^*, z^*) = P_2$ or P_3

Equation (11.3) becomes the cubic equation, with no obvious factors:

$$\lambda^3 + a_2\lambda^2 + a_1\lambda + a_0 = 0, \tag{11.4}$$

where

$$a_2 \equiv \sigma + b + 1,$$

$$a_1 \equiv b + \sigma(b + 1 - r + z^*) + x^{*2},$$

$$a_0 \equiv \sigma[b + x^{*2} + x^*y^* - b(r - z^*)].$$

They are, for either P_2 or P_3:

$$a_2 = \sigma + b + 1,$$

$$a_1 = b(r + \sigma),$$

$$a_0 = 2\sigma b(r - 1).$$

Presumably there exists a critical value of r, which we call r_c, below which the equilibrium (P_2 or P_3) is stable but above which it is unstable.

That is,

$$\text{Re}\,\lambda < 0 \quad \text{for } 1 < r < r_c,$$

$$\text{Re}\,\lambda > 0 \quad \text{for } r > r_c.$$

The boundary between stability and instability is given by

$$\text{Re}\,\lambda = 0 \quad \text{for } r = r_c.$$

To find r_c, we write

$$\lambda = i\omega$$

and assume ω to be real. Equation (11.4) becomes

$$-i\omega^3 - a_2\omega^2 + ia_1\omega + a_0 = 0.$$

Separating out the real and imaginary parts of the above equation, we get two equations:

$$-a_2\omega^2 + a_0 = 0,$$

$$-\omega^3 + a_1\omega = 0.$$

The first yields the condition

$$\omega^2 = a_0/a_2.$$

When substituted into the second, the latter becomes

$$\omega(a_0/a_2 - a_1) = 0.$$

Since ω cannot be zero (because it won't satisfy the first equation), we must have

$$a_1 a_2 = a_0 \quad \text{when } r = r_c.$$

This is a surprisingly simple result for a cubic equation:

> The product of the coefficients of the λ^2 and λ terms is equal to the constant term at the boundary where $\text{Re}\,\lambda$ is about to become positive.

Thus

$$b(r_c + \sigma)(\sigma + b + 1) = 2\sigma b(r_c - 1),$$

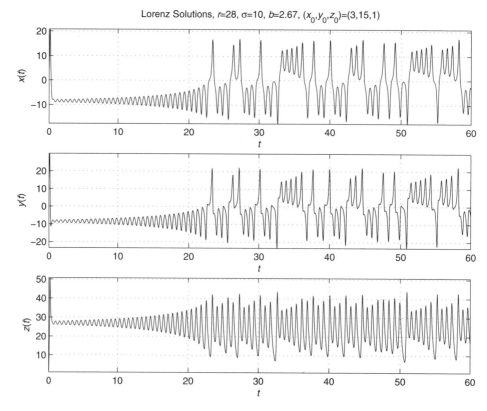

Figure 11.4. MATLAB solution of $x(t)$, $y(t)$, and $z(t)$ as a function of t for parameter values indicated.

which yields

$$r_c = \sigma \left(\frac{\sigma + b + 3}{\sigma - b - 1} \right). \tag{11.5}$$

One set of parameter values used by Lorenz was $\sigma = 10$ and $b = 8/3$. Then $r_c = 470/19 \cong 24.737$.

The Lorenz system is linearly unstable, for $r > r_c$, for small perturbations around each of its three equilibria. A nonlinear analysis of the Lorenz system is much more difficult. Lorenz showed numerically that for large r, the solution becomes chaotic. The value of r he used was $r = 28$. The numerical code (in MATLAB) is provided in Appendix B. Figure 11.4 shows each of $x(t)$, $y(t)$, and $z(t)$ as a function of t for these parameter values. The solution is chaotic.

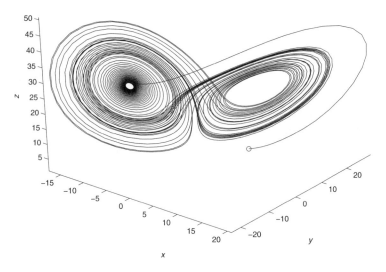

Figure 11.5. A three-dimensional plot of the trajectory $x(t)$, $y(t)$, $z(t)$ is shown. The initial point is indicated by a circle. (Drawing by Wm. Dickerson.)

A three-dimensional plot of the trajectory $(x(t), y(t), z(t))$ is shown in Figure 11.5. The famous Lorenz "butterfly" shape is seen.

The trajectories do not cross themselves in the x-y-z space, although they appear to in two-dimensional cross sections. They tend to reside on two bent surfaces (the wings of a "butterfly"), and therefore one says that their "volume" is zero (since a surface has zero volume). The trajectories appear to be attracted to the two "attractors" near P_2 and P_3. A trajectory may wind around P_2 a few times and then be repelled to the neighborhood of P_3, then wind around it a few more times before being sent back to the neighborhood of P_2, and so on. The attractors are called "strange attractors."

It can be shown that there is no limit cycle (periodic solutions) for $r > r_c$ in the Lorenz system (see exercise 7). Its solution is aperiodic (chaotic) for any value of r greater than r_c. The other possibility, that for $r > r_c$ the linearly unstable perturbations grow to infinity, can be ruled out by showing that the solution to the Lorenz system has to be finite (Lorenz, 1963). (See exercise 6.)

The nonlinear behavior of the solution for $r < r_c$ is more complicated. Although there exist two linearly stable equilibria, P_2 and P_3

(called *attracting fixed points*), there also exists *transient chaos* at finite perturbations away from these points for $r \gtrsim 13.926$. A solution not close enough to P_2, say, will be repelled towards the other fixed point, P_3, and, if it does not get close enough to P_3, gets repelled back to P_2. This would happen several times until it approaches one of the fixed points close enough to be attracted into it. On the other hand, it is easy to show (see exercise 8) that $(x^*, y^*, z^*) = (0, 0, 0)$ is nonlinearly stable ("globally attracting") for $r < 1$.

11.5 Comments on Lorenz Equations as a Model of Convection

You probably have noticed that we did not derive the Lorenz equations as a model of a physical system. This is for good reason.

Lorenz's three equations are *not* a good model of the Rayleigh–Bénard convection, although it is based on a set of two partial differential equations describing the fluid convection in two dimensions, with heating below and cooling above (Saltzman, 1962). See Figure 11.3.

Saltzman expressed the two unknowns, the streamfunction and the temperature, in the form of a double Fourier series in the two spatial coordinates, with the coefficients of the series a function of t only. These coefficients satisfy an infinite dynamical system. Lorenz (1963) took this infinite system and truncated it to only three components, $x(t)$, $y(t)$, and $z(t)$, in effect retaining only a single Fourier harmonic in each direction. Since Lord Rayleigh in 1916 found that for r slightly supercritical, i.e., slightly larger than 1, the onset of convection rolls takes the form of a single harmonic in each of the two dimensions, Lorenz felt that his three equations "may give realistic results when the Rayleigh number is slightly supercritical" but cautioned that "their solution cannot be expected to resemble those of [the original partial differential equations] when strong convection occurs, in view of the extreme truncation." Note that interest in the Lorenz system is usually in the parameter regime that gives rise to chaos, namely at very large supercritical r, when the truncated system likely fails!

When r becomes large, the other terms dropped by Lorenz—those representing smaller spatial scales and their interactions with the large scales—become important and cannot be ignored. Curry et al. (1984) looked into the problem of what happens when more and more of the harmonics of the original infinite set are retained. Chaos *disappears* when a sufficient number of terms are kept! In fact, the original partial differential equations describing convection in two dimensions do not possess chaotic behavior.

However, Lorenz's original intention was not to model convection but to look for a set of simple equations that possess the properties he

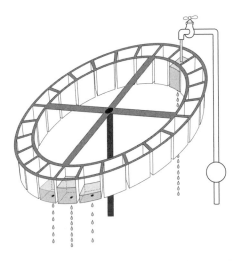

Figure 11.6. Chaotic waterwheel. (Drawing by Wm. Dickerson.)

saw in more complicated systems: aperiodicity and extreme sensitivity to initial conditions. We now know many simple *physical* systems that have the mathematical properties Lorenz sought. An example is the driven pendulum (see exercises 2 and 3 at the end of this chapter).

11.6 Chaotic Waterwheel

Another example is the leaky waterwheel of Malkus and Howard, which is discussed in Strogatz (1994). See Figure 11.6. In a "reverse" modeling effort, applied mathematicians Willem Malkus and Lou Howard at MIT constructed a mechanical contraption that possesses the chaotic properties of the Lorenz system. It is a wheel tilted from the horizontal. On its rim are attached plastic cups hanging vertically, each with a small hole of equal size at its bottom. A faucet is located at the topmost point of the wheel, dripping water into the cup there. Think of the water input rate as our parameter r. For small input rates compared to the rate at which water leaks out of the cup being filled, nothing happens. The wheel remains stationary. This corresponds to the trivial solution P_1. When the fill rate is higher than the leak rate ($r > 1$), the top cup fills up and, being top heavy, the wheel turns, either to the right or to the left, initiating either a clockwise rotation (P_2) or an anticlockwise rotation (P_3). Either rotation is stable for fixed r until r increases beyond a certain point when some of the cups get too full and are unable to make it back up to the top. The wheel slows down and may reverse direction. The wheel may spin in one direction a few times and then change direction erratically.

11.7 Exercises

1. *Damped pendulum*

A simple pendulum, with a point mass m at one end of a rigid weightless rod of fixed length L that is hinged at the other end, satisfies the following equation (Newton's second law: mass times acceleration = force):

$$m\frac{d^2}{dt^2}(L\theta) = -mg\sin\theta - m\gamma\frac{d}{dt}(L\theta).$$

Here θ is the angle the rod makes with the vertical, $L\theta$ is the displacement (arc length) from the vertical, $mg\sin\theta$ is the projection of gravitational force in the direction of motion (the angular direction), and $m\gamma\frac{d}{dt}(L\theta)$ is the air resistance, which is proportional to the velocity $\frac{d}{dt}(L\theta)$. γ is a positive constant of proportionality. This system does not possess chaotic behavior because it is a dynamical system of only second degree.

a. Let $x = \theta$, $y = \frac{d}{dt}\theta$. Write the above equation in the form of a dynamical system of the form

$$\frac{dx}{dt} = f(x, y),$$

$$\frac{dy}{dt} = g(x, y).$$

Denote g/L by ω^2.

b. Find the equilibria of the system in (a).

c. Determine the linear stability of the equilibria found in (b).

2. *Forced, damped pendulum*

Adding a forcing term to the equation in exercise 3, we have the following equation for the case of a forced damped pendulum:

$$\frac{d^2}{dt^2}\theta + \omega^2\sin\theta + \gamma\frac{d}{dt}\theta = D\cos(\omega_D t),$$

where $\omega = \sqrt{g/L}$, and ω_D is the (specified) frequency of the forcing.

Write this equation in the form of a dynamical system:

$$\frac{dx}{dt} = f(x, y, z),$$

$$\frac{dy}{dt} = g(x, y, z),$$

$$\frac{dz}{dt} = h(x, y, z),$$

thus showing that you need a system of three equations. (Define z appropriately. Note that f, g, h do not contain t explicitly.)

3.   Use a MATLAB program to plot $y \equiv \frac{d}{dt}\theta$, the angular velocity, as a function of nondimensional time ωt. Vary D and γ until you get chaotic-looking behavior.

4.   Let $\mathbf{F} = (f, g, h)$, where f, g, h are functions of x, y, and z. In addition, they are the right-hand sides of the dynamical system:

$$\frac{dx}{dt} = f(x, y, z), \frac{dy}{dt} = g(x, y, z), \frac{dz}{dt} = h(x, y, z).$$

a. Find $\nabla \cdot \mathbf{F} \equiv \frac{\partial}{\partial x} f + \frac{\partial}{\partial y} g + \frac{\partial}{\partial z} h$ for the Lorenz system. Show that it is a negative constant.

b. Let D_0 be the region in the x-y-z space where the trajectories of $x(t)$, $y(t)$, and $z(t)$ reside at $t = 0$ and let $D(t)$ be the region at time t. Let $V(t)$ be the "volume" of this region $D(t)$:

$$V(t) = \int_{D(t)} dx\, dy\, dz.$$

We wish to find the time rate of change of this "moving volume." Show that in general we have (provided that $\nabla \cdot \mathbf{F}$ exists) (*hint*: Liouville's theorem):

$$\frac{d}{dt} V(t) = \int_{D(t)} \nabla \cdot \mathbf{F}\, dx\, dy\, dz.$$

c. For the Lorenz system, show that

$$V(t) = e^{-(\sigma + b + 1)t} V(0).$$

d. What can you say about the boundedness of solutions to the Lorenz system? Does $V(t) \to 0$ as $t \to \infty$ necessarily mean that the trajectories will tend to the origin $(x, y, z) = (0, 0, 0)$?

5. Show that there are no periodic solutions of the Lorenz equations. (*Hint*: Suppose there are periodic (or quasi-periodic) solutions, whose trajectories $(x(t), y(t), z(t))$ reside on the surface of a volume $D(t)$. Then the volume should not shrink as time increases. Contrast this with the result from exercise 6.)

6. Show that for $r < 1$, the equilibrium $(x^*, y^*, z^*) = (0, 0, 0)$ is globally (nonlinearly) stable for the Lorenz system. That is, any $(x(t), y(t), z(t))$ would eventually approach $(0, 0, 0)$ as $t \to \infty$.
    Consider the "volume"

    $$V(x, y, z) = \frac{1}{\sigma}x^2 + y^2 + z^2.$$

    a. Show that, using the Lorenz equations,

    $$\frac{dV}{dt} = \frac{2}{\sigma}x\frac{d}{dt}x + 2y\frac{d}{dt}y + 2z\frac{d}{dt}z$$

    $$= -2\left[x - \frac{r+1}{2}y\right]^2 - 2\left[1 - \left(\frac{r+1}{2}\right)^2\right]y^2 - 2bz^2.$$

    b. Show that $\frac{d}{dt}V$ is strictly negative unless one reaches $(x, y, z) = (0, 0, 0)$. Thus argue that the point $(0, 0, 0)$ is the final destination of all trajectories (x, y, z) for $r < 1$.

12

El Niño and the Southern Oscillation

Mathematics introduced:
> advection equation and its finite difference approximation

12.1 Introduction

The 1997–1998 El Niño was the most severe event of its kind on record, eclipsing even the 1982–1983 "El Niño of the century." It deranged atmospheric weather patterns around the world, killed over 2,000 people, and was responsible for $33 billion in property damage worldwide. Curt Suplee, a *Washington Post* science writer, wrote in the February 1999 issue of *National Geographic*:

> It rose out of the tropical Pacific in late 1997, bearing more energy than a million Hiroshima bombs.... Peru was where it all began, but El Niño's abnormal effects on the main components of climate—sunshine, temperature, atmospheric pressure, wind, humidity, precipitation, cloud formation, and ocean currents—changed weather patterns across the equatorial Pacific and in turn around the globe. Indonesia and surrounding regions suffered months of drought. Forest fires burned furiously in Sumatra, Borneo, and Malaysia, forcing drivers to use their headlights at noon. The haze traveled thousands of miles to the west into the ordinarily sparkling air of the Maldives Islands, limiting visibility to half a mile at times. Temperature reached 108°F in Mongolia; Kenya's rainfall was 40 inches above normal; central Europe suffered a record flooding that killed 55 in Poland and 60 in the Czech Republic; and Madagascar was battered with monsoons and cyclones. In the U.S. mudslides and flashfloods flattened communities from California to Mississippi, storms pounded the Gulf Coast, and tornadoes ripped Florida.

Some of these individual disasters may just be coincidental, but El Niño and the associated abnormal floods and droughts were actually predicted by a simple climate model months in advance.

Figure 12.1. Sir Gilbert Walker. (Courtesy of Eugene M. Rasmusson, University of Maryland.)

El Niño is now defined as an anomalous warming of the surface waters of the tropical eastern Pacific Ocean—from the South American coast to the International Date Line—that persists for three or more seasons. The name actually originated in the 18th century with Spanish sea captains and, later, Peruvian fishermen, and referred to the weaker and far more benign seasonal warming of the ocean surface off the coast of Peru near Christmas—hence the name "El Niño," or "The Child."

Sir Gilbert Walker (Figure 12.1) arrived in India in 1904 as the Director General of the Observatories of India following the devastating monsoon of 1899 (an El Niño year) and the associated famine. In an attempt to predict monsoon failures in India, Walker looked for indicators and correlates all over the globe. In 1920 he discovered the "Southern Oscillation (SO)," which he defined as the pressure difference between Tahiti in the central equatorial Pacific and Darwin, Australia, in the western equatorial Pacific. Remarkably, pressure measurements from these two stations, separated by thousands of kilometers of ocean, rise and fall in a see-saw pattern. When pressure at Tahiti is high, that at Darwin is low, and vice versa. Normally, Tahiti's pressure was higher than Darwin's, but when this Southern Oscillation index became notably weak in some years, Walker found that there would be heavy rainfall in the central Pacific, drought in India, warm winters in southwestern Canada, and cold winters in the southeastern United States. Walker's work was ignored for decades until Jacob Bjerknes (Figure 12.3) advanced a conceptual model in 1969 that tied the Southern Oscillation in the atmosphere to the El Niño in the equatorial ocean. The coupled atmosphere–ocean

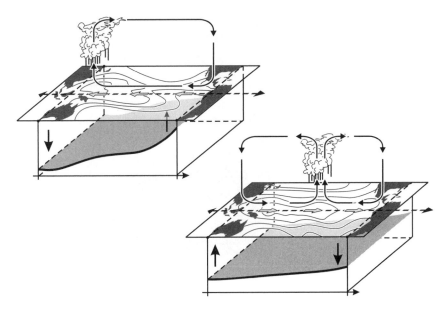

Figure 12.2. Top panel: Normal conditions in the equatorial Pacific. Lower panel: El Niño conditions. The thick line in the ocean box denotes the thermocline, separating cold deep water from the warmer ocean surface water. (Drawing by Wm. Dickerson.)

phenomenon is now referred to as El Niño–Southern Oscillation, or ENSO for short. Mark Cane and his student Stephen Zebiac constructed a simple coupled atmosphere–ocean model in the mid 1980s (Zebiac's MIT Ph.D. thesis in 1984; Zebiac and Cane, 1987). That model for the first time predicted the onset of the 1997–1998 El Niño, a remarkable achievement in climate prediction. (Recall Lorenz's conclusion that weather cannot be predicted in detail more than two weeks in advance!)

12.2 Bjerknes' Hypothesis

The waters of equatorial oceans are usually cold at depth and warm near the surface, where they are heated by the sun. The warm and cold waters are separated by a sharp transition region called the thermocline. See Figure 12.2. Under normal conditions, the trade wind over the equatorial Pacific is easterly; i.e., it blows from east to west. The easterly trade wind tends to blow the warm surface water to the western boundary of the Pacific ocean basin, where it is piled up. The sea level in the Philippines is usually 60 cm (23 inches) higher than that near Panama. The western ocean is also warmer. In fact, the western Pacific has the warmest ocean surface on earth, as high as 31.5°C (89°F). The thermocline here is about 100 to 200 m (330–660 ft) deep. Near the east coast of the ocean, the coast of Peru, the depth of the warm upper layer

Figure 12.3. Professor Jacob Bjerknes of UCLA. (Courtesy of Eugene
M. Rasmusson, University of Maryland.)

is shallower, about 40 m (130 ft) deep. Thus the cold water below the
thermocline is closer to the surface (see Figure 12.2, top). Since the deep
water is full of nutrients, the east coast of the Pacific supports a rich
variety of fish, the birds that feed on the fish, and a thriving fertilizer
industry dependent on bird droppings.

The trade winds, in turn, are driven by the sea-surface temperature
patterns. In the tropics, the region of heavy precipitation and convection in the atmosphere—with the associated ascending motion—tends
to form over the warmer waters, and sinking motion tends to occur
over the cooler ocean waters. Consequently, there is a rising motion
due to convection near Indonesia in the western Pacific and sinking
motion over the eastern Pacific near Peru. Completing the atmospheric
circulation is a surface wind from the east to the west. Bjerknes named
this atmospheric circulation pattern the *Walker circulation*, in honor of
Sir Gilbert Walker.

Bjerknes' hypothesis is that El Niño is a breakdown of these mutually
reinforcing atmosphere–ocean patterns. If for some reason the easterly
trade winds slacken, then the warm surface waters from the west would
slosh back to the east coast, depressing the thermocline there and
causing a warming of the eastern Pacific. When the east–west ocean
temperature contrast is reduced, the strength of the Walker circulation
is reduced, the trade winds weaken further and may even reverse, and
this further acts to warm the eastern Pacific (see Figure 12.2, bottom).
The warming of the eastern Pacific is called the El Niño, and the
associated atmospheric connection is the *Southern Oscillation*.

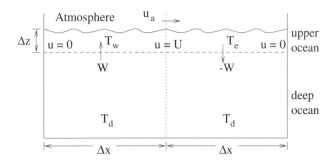

Figure 12.4. A simple model of the equatorial Pacific.

Bjerknes' model is not actually a mathematical model of ENSO but a conceptual one. Nevertheless, it has most of the ingredients for a mathematical model, but not being expressed in a mathematical form, it produces no quantitative results that can be used for verification against observation. Furthermore, the conceptual model cannot be used for prediction, as it does not explain what causes the transition from the normal state of affairs to an El Niño event and vice versa.

This "chicken and egg" problem is common to coupled oscillations. If it can be shown that the ENSO interaction *is* a coupled oscillation, then it is not necessary to enquire about an external cause for the start of an El Niño event. Nor is it necessary to attribute the eastward warming of the ocean surface to the weakening of the trade winds, or, conversely, attribute the weakening trade winds to the warming of the eastern Pacific. Furthermore, since the occurrence of ENSO is rather irregular, and probably chaotic, it would be nice if the model exhibited similar behavior in its oscillations. A simple model with all these attributes is described below. It couples the Walker circulation to the east–west sea-surface temperature difference, the latter being in turn driven by the trade winds of the Walker circulation. And, since the movement of surface waters would also involve an ocean circulation, a mass exchange with the deep water must occur.

12.3 A Simple Mathematical Model of El Niño

We discuss here a simple mathematical implementation of Bjerknes' conceptual model, by G. K. Vallis (1988). See Figure 12.4.

The Atmosphere

Let u_a be the surface wind in the atmosphere, positive if from the west. It is driven by the east–west pressure difference in the atmosphere, i.e., the SO, which is in turn driven by the east–west temperature difference, convection, and the Walker circulation. In this simple model, it is assumed that it is driven by the east–west temperature difference, $(T_e - T_w)$, and that it will relax back to some "normal" equatorial easterlies ($u_0 \leq 0$, specified from observation of "normal" conditions) in the absence of the temperature difference. Thus:

$$\frac{du_a}{dt} = b(T_e - T_w) + r(u_0 - u_a), \qquad (12.1)$$

where r is the rate at which u_a relaxes to u_0 in the absence of coupling with the ocean temperature. The parameter b is the rate at which the ocean is influencing the atmosphere.

Since the ocean changes slowly and the atmosphere responds to oceanic changes rapidly, it is a good assumption that the atmosphere is in "quasi-equilibrium" with the ocean. Thus, at the time scale appropriate to the ocean, $\frac{d}{dt}u_a \sim 0$ (i.e., the atmosphere has already reached a steady state given an ocean temperature because the latter changes slowly):

$$0 \cong b(T_e - T_w) - r(u_a - u_0).$$

Solving for u_a:

$$u_a = \frac{b}{r}(T_e - T_w) + u_0. \qquad (12.2)$$

So the atmosphere is "known" if the ocean temperature is known.

Air–Sea Interaction

Let U be the surface ocean current at midbasin; it is dragged by the wind stress at the air–sea interface. We assume that it is driven by the surface wind u_a:

$$\frac{dU}{dt} = Du_a - CU. \qquad (12.3)$$

In the absence of surface wind, U will relax to 0 with a time scale of C^{-1}.

Substituting u_a from (12.2) into the above equation, we get an ordinary differential equation for the ocean current U:

$$\frac{dU}{dt} = B(T_e - T_w) - C(U - U_0),$$

(12.4)

where $B \equiv Db/r$, $U_0 \equiv Du_0/C$.

Ocean Temperature Advection

The equation describing the change in temperature due to advection of a warmer or colder temperature from elsewhere by a current u in the x-direction and a current w in the z-direction is (see the appendix to this chapter, section 12.5, for a derivation of the advection equation):

$$\frac{\partial T}{\partial t} + u\frac{\partial}{\partial x}T + w\frac{\partial}{\partial z}T = 0.$$

We use a very crude finite difference scheme to turn this partial differential equation into ordinary differential equations.

Let W be the vertical velocity in the western box across the thermocline. To conserve mass flux across the thermocline, it must be $-W$ in the eastern box (see Figure 12.4).

Consider the eastern box:

$$\frac{\partial T_e}{\partial t} + U\frac{T_e - T_w}{\Delta x} + (-W)\frac{T_e - T_d}{\Delta z} = 0;$$

and the western box:

$$\frac{\partial}{\partial t}T_w + U\frac{T_e - T_w}{\Delta x} + W\frac{T_w - T_d}{\Delta z} = 0.$$

(Recall that T_d is the temperature of the deep water; see Figure 12.4.) Mass conservation considerations relate U and W. The density of sea water is considered as approximately constant. The mass flux is the constant density times the velocity. Conservation of mass then implies $W\Delta x = U\Delta z$, or

$$W = U\Delta z/\Delta x.$$

The temperature advection equation then becomes

$$\frac{\partial T_e}{\partial t} = \frac{U}{\Delta x}(T_w - T_d),$$

$$\frac{\partial T_w}{\partial t} = \frac{U}{\Delta x}(T_d - T_e).$$

To the right-hand sides we add a relation to a prescribed temperature T_0 in the absence of temperature advection. That is, in the absence of an east–west temperature difference, the above equation will not determine the ocean surface temperature. That temperature is supposedly determined by processes not in our model. A simple way to deal with this problem is to specify that temperature T_0 to which the temperatures will relax with a rate A. T_0 is determined by heat transfers (radiative and diffusive) from the upper ocean to the atmosphere and to the deep ocean.

$$\frac{\partial T_e}{\partial t} = \frac{U}{\Delta x}(T_w - T_d) - A(T_e - T_0),$$

$$\frac{\partial T_w}{\partial t} = \frac{U}{\Delta x}(T_d - T_e) - A(T_w - T_0).$$

(12.5)

Without loss of generality we set $T_d = 0$ (we can always measure temperature relative to that of the deep water). The equations are made dimensionless by letting

$$x = U/(A\Delta x),$$

$$y = (T_e - T_w)/(2T_0),$$

$$z = 1 - (T_e + T_w)/(2T_0),$$

$$\hat{t} = At.$$

Then

$$\text{ENSO} \begin{cases} \dfrac{dx}{d\hat{t}} = \sigma y - \rho(x - x_0), \\[2mm] \dfrac{dy}{d\hat{t}} = x - xz - y, \\[2mm] \dfrac{dz}{d\hat{t}} = xy - z, \end{cases}$$

(12.6)

where

$$x_0 = U_0/(A\Delta x), \sigma = 2BT_0/(\Delta x A^2), \text{ and } \rho = C/A.$$

This set is comparable to the Lorenz equations if $x_0 = 0$:

$$\text{Lorenz} \begin{cases} \dfrac{dx}{dt} = -\sigma x + \sigma y, \\[2mm] \dfrac{dy}{dt} = rx - xz - y, \\[2mm] \dfrac{dz}{dt} = xy - bz. \end{cases} \tag{12.7}$$

In fact, the ENSO set becomes the Lorenz set under the following rescaling (for the case $x_0 = 0$):

$$x = \alpha\hat{x}, \quad y = \beta\hat{y}, \quad z = \zeta\hat{z},$$

$$\alpha\zeta/\beta = 1, \quad \beta/\alpha = \rho/\sigma, \quad \alpha = 1,$$

$$\text{rescaled ENSO} \begin{cases} \dfrac{d\hat{x}}{d\hat{t}} = \rho\hat{y} - \rho\hat{x}, \\[2mm] \dfrac{d\hat{y}}{d\hat{t}} = (\sigma/\rho)\hat{x} - \hat{x}\hat{z} - \hat{y}, \\[2mm] \dfrac{d\hat{z}}{d\hat{t}} = \hat{x}\hat{y} - \hat{z}. \end{cases}$$

We therefore identify σ/ρ with Lorenz's r and ρ with Lorenz's σ and set Lorenz's b to 1. We will, however, discuss the results below using the original (unscaled) variables x, y, z, which are more physical.

The case of $x_0 = 0$ will be considered first along the lines used previously to study the Lorenz equations. Just as in the Lorenz equations, $(x, y, z) = (0, 0, 0)$ could be the solution. This would have implied that there is no coupled atmosphere–ocean oscillation, and thus no ENSO. It is possible, however, that the parameters in the equatorial Pacific are such that the trivial solution is not attainable because it is unstable. We will look into the stability issue next.

Using analogy with the Lorenz equations, we see that for $r \equiv \sigma/\rho < 1$, the trivial solution

$$P_1 = (x_1^*, y_1^*, z_1^*) = (0, 0, 0)$$

is the only realizable equilibrium. Physically, this solution represents the case when there is no east–west temperature difference in the ocean, no ocean advection, no trade wind, and the surface temperature in the ocean equals T_0.

For $r \equiv \sigma/\rho > 1$, there are three equilibria—P_1, plus two new solutions:

$$P_2 = \left((r-1)^{1/2}, \ \left(1/r \left(1 - \frac{1}{r} \right) \right)^{1/2}, \ 1 - 1/r \right),$$

$$P_3 = \left(-(r-1)^{1/2}, \ -\left(\frac{1}{r}(1 - 1/r) \right)^{1/2}, \ 1 - 1/r \right).$$

P_1 becomes unstable. P_2 and P_3 are stable until $r > r_c$. From Lorenz equation results in chapter 11, all three equilibria lose stability when

$$\sigma > \sigma_c \equiv \frac{(4 + \rho)\rho^2}{\rho - 2}.$$

For the Pacific Ocean, $\Delta x = 7{,}500$ km, $T_0 \sim 15°$ (as measured from T_d), the frictional decay rate $C \sim 1/(2 \text{ months})$, and the temperature decay rate is $A \sim 1/(6 \text{ months})$. If $B\Delta x \sim 12 \text{m}^2\text{s}^{-2}C^{-1}T_0^{-1}$, then $\rho \sim 3$, $\sigma_c \sim 63$, and $\sigma \sim 102$.

Therefore the system could very well be in the unstable regime. Drawing upon what we know about the Lorenz system, which possesses irregular oscillations when all three of its equilibria become unstable, we anticipate that Vallis's system also produces irregular oscillations ranging from El Niño, to normal, to perhaps La Niña (anti–El Niño) events. Whether such oscillations bear any resemblance to the observed ones can be revealed by displaying the solution numerically, using MATLAB. The MATLAB code is given in Appendix B.

Figure 12.5 displays the east–west temperature difference (in terms of $y(t) \equiv (T_e - T_w)/2T_0$) as a function of time (in terms of $\hat{t} = At$). Since A, the thermal relaxation rate of the ocean, is taken to be $1/(6 \text{ months})$, each two units of time plotted correspond to one year. The top panel of Figure 12.5 shows $y(t)$ for 30 years for the case of $U_0 = 0$. The bottom panel is for $U_0 = -0.45$ m/s. Both figures show an irregular oscillation, which is self-sustained (i.e., without the need for external anomalous forcing). Without a prevailing easterly pushing the warm water westward, it is equally likely to have El Niño (the extreme warming of the eastern Pacific) as La Niña (the extreme cooling of the eastern Pacific). This symmetry is broken when there is a prevailing easterly, as shown in the bottom panel of Figure 12.5. In this case, on average, T_e is colder than T_w; hence the average of $y(t)$ in time is negative. El Niño is the large positive deviation from the time average, while La Niña is the large negative deviation. El Niño is more frequent than La Niña. Furthermore, the interval between El Niño events becomes longer, and closer to the observed situation, when there is a prevailing easterly wind.

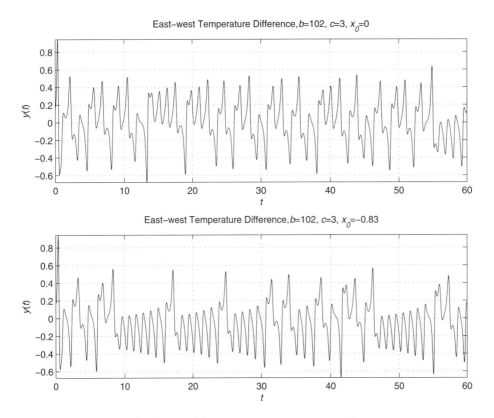

Figure 12.5. Numerical solution of the east–west temperature difference as a function of time.

12.4 Other Models of El Niño

The Vallis model presented in the previous section is not the type of model used for prediction or even quantitative analysis of ENSO. For one thing, the finite difference adopted is too coarse ($\Delta x = 7500$ km!). Also, as a result of coarse resolution, waves, which propagate signals from one ocean boundary to the other, are not resolved. The only effect incorporated is that of temperature advection by ocean currents.

These equatorial ocean waves were incorporated in the original model of Zebiac and Cane (1987), involving coupled partial differential equations. Recently there has been an attempt to model the ENSO oscillations as a delayed oscillator, taking into account the fact that the signal travels at the speed of the equatorial ocean waves.

The idea that ENSO is a self-sustaining chaotic system of relatively low order has been challenged by Penland and Sardeshmukh (1995), Thompson and Battisti (2000, 2001), and others who have shown that rapidly varying forcing (in surface wind stress and heat flux) could be the source of the irregularity of El Niño. Models involving "stochastic" (random) high-frequency forcing are being studied, even using linear models.

12.5 Appendix: The Advection Equation

A useful conservation equation is the advection equation, which describes the evolution in space and time of a quantity, F, say, which is *conserved*. That is, F does not change if the observer is moving with it; i.e.,

$$\frac{dF}{dt} = 0.$$

The coordinate system $(x(t), y(t), z(t))$, which is attached to the observer, is moving with the velocity

$$(u, v, w) = \left(\frac{dx}{dt}, \frac{dy}{dt}, \frac{dz}{dt} \right)$$

relative to a stationary frame of reference. Such a moving coordinate system is called the Lagrangian coordinates. Newtonian mechanics is usually expressed in terms of Lagrangian coordinates. The second law of motion, e.g., is expressed as

$$\frac{d}{dt}(mv) = f.$$

That is, the rate of change of momentum (mv) *following the particle* of mass m is equal to the applied force f.

Often, however, we desire to write the equations in a stationary frame of reference, called the "laboratory" frame. This frame is adopted in the field of fluid dynamics, where it is almost impossible to track every fluid particle with its own Lagrangian coordinate system. Instead, the observer simply sits in a stationary frame and lets the particles move by with velocities (u, v, w). Since the velocities and the quantity F are different at different locations and times, they are in turn functions of x, y, z, and t, where x, y, z are now fixed. This coordinate system is called the Eulerian coordinates.

We now need a transformation from a Lagrangian coordinate description to a Eulerian description:

$$\frac{d}{dt} F(x(t), y(t), z(t), t)$$

$$= \frac{dx}{dt} \frac{\partial}{\partial x} F + \frac{dy}{dt} \frac{\partial}{\partial y} F + \frac{dz}{dt} \frac{\partial}{\partial z} F + \frac{\partial}{\partial t} F.$$

The partial derivative with respect to x, $\frac{\partial}{\partial x}F$, means that the derivative is taken while holding all other variables (in this case, y, z, and t) fixed. Similarly, $\frac{\partial}{\partial t}F$ means holding x, y, z fixed, even though x, y, z also depend on t in the Lagrangian description.

Since

$$(u, v, w) = \left(\frac{dx}{dt}, \frac{dy}{dt}, \frac{dz}{dt} \right),$$

we have

$$\frac{d}{dt}F = u\frac{\partial}{\partial x}F + v\frac{\partial}{\partial y}F + w\frac{\partial}{\partial z}F + \frac{\partial}{\partial t}F.$$

The final step in the transformation is to treat F from now on as functions of fixed space x, y, z, and time t. Thus, the advection equation becomes

$$\left(\frac{\partial}{\partial t} + u\frac{\partial}{\partial x} + v\frac{\partial}{\partial y} + w\frac{\partial}{\partial z} \right) F(x, y, z, t) = 0.$$

12.6 Exercises

1.   One of the criticisms of the simple three-component model of ENSO presented here is the coarse resolution of the finite difference adopted for x and z derivatives in the ocean advection part of the model. The result turns out to be qualitatively sensitive to the particular type of finite difference scheme used, as pointed out by the author, Vallis, himself. In section 12.3, we derived a Lorenz-type system by using a centered differencing scheme. An equally reasonable scheme, no more or less accurate, is the upstream scheme. Under this scheme, Eq. (12.5) is replaced by

$$\frac{\partial}{\partial t}T_e = \frac{U}{\Delta x}(T_w - T_e) - A(T_e - T_0) \quad \text{if } U > 0,$$

$$\frac{\partial}{\partial t}T_e = \frac{U}{\Delta x}(T_e - T_d) - A(T_e - T_0) \quad \text{if } U < 0,$$

$$\frac{\partial T_w}{\partial t} = \frac{U}{\Delta x}(T_d - T_w) - A(T_w - T_0) \quad \text{if } U > 0,$$

$$\frac{\partial T_w}{\partial t} = \frac{U}{\Delta x}(T_w - T_e) - A(T_w - T_0) \quad \text{if } U < 0.$$

The U equation remains as in Eq. (12.4), but we will consider only the case of $U_0 \equiv 0$.

a. Show that there is an east–west symmetry. That is, reversing the sign of U and substituting T_w for T_e and T_e for T_w leads to an identical equation set.

b. Because of (a), we need to consider only one sign of U. Take $U > 0$. Define x, y, z as was done in section 12.3. Write down the system of three ordinary differential equations for these variables.

c. Find the equilibria.

d. Determine the linear stability of each equilibrium.

e. Discuss the possible steady states of this coupled atmosphere–ocean model.

f. Can you show that there is no limit cycle solution in this model?

13

Age of the Earth: Lord Kelvin's Model

Mathematics required:
> solution of first-order ordinary differential equation by
> separation of variables; multivariate calculus; partial derivatives

Mathematics introduced:
> the method of similarity solution in solving certain partial
> differential equations

13.1 Introduction

This subject is full of controversy, some of it scientific and some of it religious. Our interest here is not to settle the controversy or to take sides in the debate. Rather, we are interested in framing a mathematical problem in the context of the debate and seeing how modeling can be used to add quantitative information. The discovery of radioactivity and the development of radiometric dating methods are quite recent (in the early and mid 20th century, respectively). Imagine if you were born earlier, say in the 18th century, or that you don't believe in the validity of the current radiometric dating methods, as some people still don't. How do you arrive at an estimate of the age of the earth? This was the task facing William Thomson, Lord Kelvin (1824–1907) (Figure 13.1).

The prevailing views have swung pendulum-like between different extremes in the past few centuries. The current *scientific* view is that the earth is about 4.5 billion years old. Radiometric dating of rocks on earth found ancient rocks of about 3.5 billion years old, with the oldest rock found so far, in Northwestern Canada near Great Slave Lake, dating to 3.96 billion years. In Western Australia, geologists found a zircon crystal 4.4 billion years old trapped inside a rock dated 3.1 billion years old. (An article in the September 2001 issue of *National Geographic* has an interesting description of the zircon crystals, found when magma cools, trapping a few uranium atoms in their lattices. The uranium atoms are protected from outside contaminants for billions of years: "Zircons are God's gift to geochemistry.") Moon rocks returned to Earth by the six Apollo and three Luna missions have been dated at 4.4–4.5 billion years. The meteorites have been dated at 4.58 billion years.

Figure 13.1. Lord Kelvin (William Thomson) (1824–1907).

Even today, Young Earth advocates, such as ICR (the Institute for Creation Research), believe in a literal reading of the Genesis account. Each day was a 24-hour day comparable to a modern day, and plants and animals were created in a mature functional state directly by God. By examining the various genealogies found in the Scriptures, it was estimated that Creation must have taken place sometime between six and fifteen thousand years ago. The most famous biblical chronology is due to Archbishop James Ussher, Primate of All Ireland, Vice Chancellor of Trinity College in Dublin, who in 1650 determined the date of Creation to be Sunday, October 23, 4004 BC, precisely! Modern creationists are willing to acknowledge the possibility of gaps in the genealogies, pushing this date back some. Even within the creationist camp there is currently a debate between the "Young Earth" creationists and the "Old Earth" creationists. The Old Earth creationists do not take the length of a day literally as 24 hours ("With the Lord a day is like a thousand years, and a thousand years are like a day," 2 Peter 3:8). However, as pointed out by the other camp, that Genesis intended a day to be 24 hours is made clear from the phrase following the description of each day: "There was evening, and there was morning." To make it even more explicit, the terms used are defined in Genesis 1:5, where "God called the light 'day' and the darkness He called 'night'." It is also pointed out by some in the Young Earth camp that the creation events are not in the expected order if long periods of time were involved: plants were created on day three, while the sun was not created until day four, and animals needed for the pollination of plants were not available until day five.

In Lord Kelvin's time (1850s) the debate was between the "Doctrine of Uniformity" and others such as Kelvin's. In *Proceedings of the Royal Society of Edinburgh* Lord Kelvin wrote:

> The Doctrine of Uniformity in Geology, as held by many of the most eminent of British geologists, assumes that the earth's surface and upper crust have been nearly as they are at present in temperature and other physical qualities during millions of millions of years. But the heat which we know, by observation, to be now conducted out of the earth yearly is so great, that if *this* action had been going on with any approach to uniformity for 20,000 million years, the amount of heat lost out of the earth would have been about as much as would heat, by 100° cent., a quantity of ordinary surface rock of 100 times the earth's bulk. This would be more than enough to melt a mass of surface rock equal to the bulk of the *whole earth*. (Thomson, 1866, pp. 512–513)

13.2 The Heat Conduction Problem

So, if not for 20,000 million years, how long? The problem from the perspective of heat conduction was solved by Thomson. At the age of 16 in 1840, at the University of Glasgow, Thomson read Fourier's *The Analytical Theory of Heat*. He later said: "I took *Fourier* out of the University Library, and in a fortnight I had mastered it—gone right through it." The subject of heat became a lifelong interest. The Kelvin temperature scale was later named after his work on absolute temperature in 1848.

In 1864, Thomson produced an estimate of the age of the earth by considering it as a warm, chemically inert sphere cooling through its surface. Thomson knew at the time that the temperature of the earth is hotter within and that the rocks were molten. Assuming that the hotter temperature was the temperature of the earth at an earlier time, he attempted to deduce the age of the earth. The mathematical formulation is as follows.

Let $t = 0$ be the time (the "beginning") when the earth's surface first solidified. The temperature of the earth at the time, $u_0(y)$, is taken to be the melting temperature of rock (which can be measured). y is the depth from the surface. He also assumed that the temperature at the surface of the earth for $t > 0$ has been more or less constant, i.e., $u(0, t) = u_s, t > 0$, and so u_s can be measured at present. This temperature is maintained against heat lost to space by heat conducted from below the surface. The equation of heat conduction (discovered by Joseph Fourier, 1800) was known to Thomson:

$$\frac{\partial}{\partial t} u = \alpha^2 \frac{\partial^2}{\partial y^2} u, \quad y > 0, \quad t > 0,$$

(13.1)

where $u(y, t)$ is the temperature at depth y and time t. α^2 is the conductivity of the earth and can be determined by measuring samples of surface rock. We have:

partial differential equation (PDE): $\quad \frac{\partial}{\partial t} u = \alpha^2 \frac{\partial^2}{\partial y^2} u, \; y > 0, \; t > 0,$

boundary condition (BC): $\quad u(0, t) = u_s, \; t > 0,$

initial condition (IC): $\quad u(y, 0) = u_0, \; y > 0.$

We can make the boundary condition homogeneous by letting

$$\psi(y, t) \equiv u(y, t) - u_s.$$

So in terms of ψ, the problem is:

PDE : $\quad \frac{\partial}{\partial t} \psi = \alpha^2 \frac{\partial^2}{\partial y^2} \psi, \; 0 < y < \infty, \; t > 0,$

BC : $\quad \psi(0, t) = 0, \; t > 0,$

IC : $\quad \psi(y, 0) = \psi_1 \equiv u_0 - u_s, \; y > 0.$

(13.2)

Lord Kelvin actually treated the earth as a sphere. We have simplified the problem to one dimension since we are interested only in the variation with depth. We will also artificially extend the domain to $0 < y < \infty$ for mathematical convenience. This does not cause any problem since the influence of the surface decays rapidly with depth. The fact that the mantle and the core of the earth are not solid also does not matter for the same reason.

Solution

In this problem, y measures depth and therefore has the dimension of length (in either meters or feet), t is time (in units of seconds), and the coefficient of conductivity α^2 is typically $0.012 \text{ cm}^2/\text{s}$. ψ is a temperature in either degrees Celsius or degrees Fahrenheit.

To solve this problem, we assume that $\psi(y, t)$ depends on y and t in the following combination:

$$\psi(y, t) = F\left(\frac{y}{\sqrt{\alpha^2 t}}\right). \tag{13.3}$$

This is a consequence of the fact that $y/\sqrt{\alpha^2 t}$ is a dimensionless quantity. That is, all units cancel in this combination (check this!). It turns out that it is the only possible combination of quantities in the problem that will make either y or t dimensionless. Since all mathematical functions must involve a dimensionless argument (e.g., $\sin(\omega t)$, $\exp(rt)$), it then follows that $\psi(y, t)$ must depend on y and t in the combination assumed. Had there been a spatial scale L and a time scale T, the solution could have been written as $F(y/L, t/T)$, and we would not have made much progress. In our case we let:

$$z \equiv \frac{y}{\sqrt{\alpha^2 t}};$$

then

$$\psi(y, t) = F(z).$$

This assumption is called a *similarity assumption* and works sometimes for problems without a typical length and a typical time scale. Assuming this is true, then $(F'(z) \equiv \frac{d}{dz} F(z))$

$$\frac{\partial}{\partial t} \psi = \frac{\partial z}{\partial t} \cdot \frac{d}{dz} F(z) = \frac{-\frac{1}{2}y}{\sqrt{\alpha^2 t}} \frac{1}{t} F'(z) = -\frac{1}{2} z \frac{1}{t} F'(z),$$

$$\frac{\partial}{\partial y} \psi = \frac{\partial z}{\partial y} \frac{d}{dz} F(z) = \frac{1}{\sqrt{\alpha^2 t}} F'(z),$$

$$\frac{\partial^2}{\partial y^2} \psi = \frac{1}{\alpha^2 t} F''(z), \quad \alpha^2 \frac{\partial^2}{\partial y^2} \psi = \frac{1}{t} F''(z).$$

Thus

$$\frac{\partial}{\partial t} \psi = \alpha^2 \frac{\partial^2}{\partial y^2} \psi$$

becomes

$$\frac{1}{t}F''(z) = -\frac{1}{2}z\frac{1}{t}F'(z),$$

which is

$$F''(z) + \frac{1}{2}zF'(z) = 0.$$

This is actually a first-order ordinary differential equation for $W(z) \equiv F'(z)$:

$$\frac{d}{dz}W + \frac{z}{2}W = 0.$$

Separation of variables yields

$$\frac{dW}{W} = -\frac{z}{2}dz.$$

So

$$W(z) = W(0)e^{-z^2/4}$$

and

$$F(z) = \int_0^z W(z)dz = W(0)\int_0^z e^{-z^2/4}dz = 2W(0)\int_0^{\frac{z}{2}} e^{-z'^2}dz'.$$

We have used the fact $F(0) = 0$ because $z = \frac{y}{\sqrt{\alpha^2 t}}$, and the boundary condition $\psi(0, t) = 0$. The initial condition $\psi(y, 0) = \psi_1$ implies that $F(\infty) = \psi_1$ since $z = \frac{y}{\sqrt{\alpha^2 t}} \to \infty$ for $t \to 0^+$, $y > 0$. Using the integral identity,

$$\int_{-\infty}^{\infty} e^{-z'^2}dz' = 2\int_0^{\infty} e^{-z'^2}dz' = \sqrt{\pi}, \quad \text{so} \quad \int_0^{\infty} e^{-z'^2}dz' = \frac{\sqrt{\pi}}{2}.$$

The condition

$$F(\infty) = \psi_1$$

then leads to

$$2W(0)\int_0^{\infty} e^{-z'^2}dz' = \psi_1.$$

So $W(0) = \psi_1/\sqrt{\pi}$. Finally

$$\psi(y, t) = \frac{2\psi_1}{\sqrt{\pi}} \int_0^{\frac{y}{2\sqrt{\alpha^2 t}}} e^{-z'^2} dz',$$

$$u(y, t) = u_s + \frac{2(u_0 - u_s)}{\sqrt{\pi}} \int_0^{\frac{y}{2\sqrt{\alpha^2 t}}} e^{-z'^2} dz'.$$

The temperature gradient is obtained as

$$\frac{\partial u}{\partial y}(y, t) = \frac{(u_0 - u_s)}{\sqrt{\pi \alpha^2 t}} e^{-\frac{y^2}{4\alpha^2 t}}. \tag{13.4}$$

The temperature gradient, $\frac{\partial}{\partial y} u(y, t)$, can be measured near the surface (but not so close to the surface as to be affected by the seasonal change of weather).

Let

$$\Delta \equiv \frac{\partial}{\partial y} u(y, t), \quad y \sim 0;$$

then

$$\boxed{\Delta = \frac{(u_0 - u_s)}{\sqrt{\pi \alpha^2 t}}.} \tag{13.5}$$

Solve for t, the age of the Earth:

$$\boxed{t = (u_0 - u_s)^2/(\pi \alpha^2 \Delta^2).} \tag{13.6}$$

This result can be anticipated from "back of envelope" dimensional arguments.

The temperature gradient is approximately equal to the temperature difference $(u_0 - u_s)$, divided by the length scale over which the difference occurs. There is no length scale in the problem other than the combination $X \equiv \sqrt{\alpha^2 t}$, which has the dimension of length. So

$$\Delta \sim \frac{(u_0 - u_s)}{\sqrt{\alpha^2 t}}.$$

Comparing this with the exact solution we see that we have missed only by a factor $\frac{1}{\sqrt{\pi}}$—not much for this problem.

13.3 Numbers

> When you cannot measure it, when you cannot express it in numbers, your knowledge is of a meager and unsatisfactory kind.
>
> — *Lord Kelvin*

Estimate

$$t = \frac{1}{\pi \alpha^2} \left(\frac{u_0 - u_s}{\Delta} \right)^2 .$$

Kelvin knew at the time that the temperature increases by about "1°F every 50 ft downward." The conductivity of the rock (Edinburgh rocks vs. Greenwich rocks) is about 0.012 cm^2/s. Kelvin gave wild guesses of the temperature to melt rocks: 10,000°F ("not realistic") and 7,000°F ("closer to the truth").

With $u_0 \sim 7{,}000°$F, Kelvin got $t = 98$ million years, with a possible range of 20 to 400 million years:

$$t \sim \frac{1}{\pi \, 0.012 \, \text{cm}^2/\text{s}} \left(\frac{7{,}000°\text{F}}{1°\text{F}/50 \, \text{ft}} \right)^2$$

$$\sim \frac{350{,}000^2 \, \text{ft}^2}{\pi \, 0.012 \, \text{cm}^2} \cdot \text{s}$$

$$= 98 \text{ million years}.$$

Kelvin also estimated the age of the sun from the solar constant, which yields the rate at which the sun is losing energy:

$$S = 1 \times 10^6 \, \text{cal} \, \text{cm}^{-2} \, \text{year}^{-1},$$

which gives the total output of the sun, $\sim 3 \times 10^{33}$ cal/year. Kelvin was unaware of the radioactive heat source. He assumed that the sun produced its energy from the gravitational potential of matter falling into the sun, including meteorites and even planets (and later, the matter that composed the mass of the sun itself), and published a value of 50 million years for the age of the sun in 1853.

Thus there seemed to be some consistency in Kelvin's estimates of the ages of the earth and the sun. But both are too short for Darwin's theory of evolution. Kelvin became an opponent of Darwin's theory.

The discovery of radioactivity at the turn of the 20th century offered a solution to the problems of both the Earth and the sun. Lord Rayleigh (see Strutt 1906) worked on the radioactivity in igneous rocks and proposed a shallow distribution of heat sources that neatly removed the thermal conductivity problem that had led Kelvin to his erroneous conclusion.

13.4 Exercises

1. Fourier's wine cellar

Joseph Fourier (1768–1830) was an expert on heat conduction. To find the perfect depth to build his wine cellar, he solved the following problem on subsoil temperature variations:

$$\frac{\partial}{\partial t}u = \alpha^2 \frac{\partial^2}{\partial y^2}u, \ \ 0 < y < \infty,$$

subject to the boundary conditions:

$$u(y, t) \text{ bounded as } y \to \infty,$$

$$u(0, t) = u_0 e^{i\omega_0 t}.$$

Here ω_0 is the frequency of the temperature variation at the surface, $\omega_0 = 2\pi/(1 \text{ day})$ for daily variations and $\omega_0 = 2\pi/(1 \text{ year})$ for seasonal variations. u is the temperature of the soil. y is the depth underground, positive downward, with $y = 0$ being the surface. For soil, the thermal conductivity is $\alpha^2 \cong 0.01 \text{ cm}^2/\text{s}$. We use the complex exponential instead of $\cos \omega_0 t$ because it makes the algebra easier. One can always take the real part of the solution afterwards.

a. Solve the problem (i.e., find $u(y, t)$) by assuming that your solution can be written in the form

$$u(y, t) = \phi(y)e^{i\omega_0 t}.$$

b. For daily variations, $u_0 = 5°C$ and $\omega_0 = 2\pi/(1 \text{ day})$. Find the depth below which the magnitude of temperature variation is less than $2°C$.

c. Do the same for seasonal variations, with $u_0 = 15°C$ and $\omega_0 = 2\pi/$ (1 year).

d. An ideal depth to locate a wine cellar is where the temperature variation is not only small but perfectly out of phase with the seasonal oscillation at the surface. At this depth it would be winter when it is summer at the surface. Find this depth. ($\sqrt{-1} = i$, $\sqrt{i} = \frac{1}{\sqrt{2}}(1+i)$, $e^{i\theta} = \cos\theta + i\sin\theta$, real part of $e^{i\theta} = \cos\theta$.)

2. **Preventing nuclear meltdown**

When the temperature in the core of a nuclear reactor reaches its critical value u_c, a meltdown occurs as the temperature from nuclear fission increases rapidly. There is very little that can be done once this happens. To protect against meltdowns, supercooled cooling rods are activated at the first sign of trouble. These rods attempt to maintain the temperatures at the two ends of a reactor (at $x = -L$ and $x = L$) at a fixed temperature u_0, which is below u_c. A one-dimensional problem for the temperature $u(x, t)$ of the nuclear reactor is given by

$$\text{PDE:} \quad \frac{\partial}{\partial t}u = \alpha^2 \frac{\partial^2}{\partial x^2}u + ae^u,$$

$$\text{BC:} \quad u = u_0 \text{ at } x = -L \quad \text{and} \quad x = L, \quad t > 0.$$

a. Find the steady state solution. (*Hint*: Set $\frac{\partial}{\partial t}u = 0$. Solve $\alpha^2 \frac{d^2}{dx^2}u + ae^u = 0$ by multiplying it by $\frac{d}{dx}u$ and integrating it in x.)
   Express the constant of integration in terms of the maximum temperature u_{max}. (*Hint*: $u = u_{max}$ where $\frac{d}{dx}u = 0$.)

b. Show that this maximum temperature occurs at the center ($x = 0$) of the reactor.

c. We want to keep the maximum temperature of the reactor safely below the critical temperature, i.e.,

$$u_{max} = u_c - b,$$

by a given margin b. What should the temperature of the cooling rods be?

d. Suppose we are too late in inserting the cooling rods and we fail to maintain the maximum temperature below the critical value. Temperature increases rapidly in time, and heat conduction is ineffective.

Dropping the heat conduction term, we have

$$\frac{\partial}{\partial t}u = ae^u.$$

Solve this equation (which is an ordinary differential equation in time) using separation of variables for ordinary differential equations. Assume u at $t = 0$ to be u_c.

14

Collapsing Bridges: Broughton and Tacoma Narrows

Mathematics introduced:
the wave equation and its solution; the concept of resonance in the context of a forced partial differential equation

14.1 Introduction

We wish to model the oscillations of suspension bridges under forcing. The forcing could come from wind, as in the case of the collapse of the Tacoma Narrows Bridge in 1940, or as a result of a column of soldiers marching in cadence over a bridge, as in the collapse of the Broughton Bridge near Manchester, England, in 1831.

These disasters have often been cited in textbooks on ordinary differential equations as examples of *resonance*, which happens when the frequency of forcing matches the natural frequency of oscillation of the bridge, with no discussion given on how the natural frequency is determined, or even where the ordinary differential equation used to model this phenomenon comes from. The modeling of bridge vibration by a partial differential equation, although still simple-minded, is a big step forward in connecting to reality.

14.2 Marching Soldiers on a Bridge: A Simple Model

In 1831, a column of soldiers marched in cadence over the Broughton Bridge near Manchester. The suspension bridge moved up and down so violently that a pin anchoring the bridge came loose and the bridge collapsed. Whether the bridge's collapse was caused by the synchronized steps of the marching or was simply a result of weight overload is not clear. This incident nevertheless led from that time on to the order to "break steps" when soldiers approach bridges. We study here the possibility that the periodic forcing may lead to resonance.

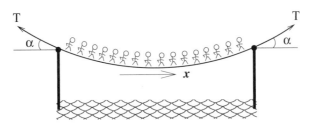

Figure 14.1. A schematic of our simple suspension bridge. (Drawing by E. Hinkle.)

When a column of soldiers marches in unison over a bridge, a vertical force

$$f(x, t)$$

is exerted on the bridge that is periodic in time t, with a period P determined by the time interval between steps (Figure 14.1). In this one-dimensional problem, with x measured along the length of the bridge, we do not distinguish left-foot steps from right-foot steps. In reality, these left–right steps create additional torsional vibrations over the width of the bridge, which, in some cases, may be more important in causing collapse. This aspect of the problem can be handled by introducing another space dimension into the model but will be ignored here.

Specifically, we will model the bridge as an elastic string of length L, suspended at only $x = 0$ and $x = L$. (We know, of course, that bridges do not behave like guitar strings. Nevertheless, this simplification allows us to skip most of the structural mechanics that one needs to know and yet still retain most of the ingredients we need to illustrate the mathematical problem of resonance.) We consider the vertical displacement $u(x, t)$ of the string (i.e., bridge) from its equilibrium position, where x is the distance from the left suspension point, and t is time. We consider a small section of the string between x and $x + \Delta x$. See Figure 14.2.

We apply Newton's law of motion,

$$ma = F$$

(mass times acceleration balancing force), to the vertical motion of this small section of the string. Its mass m is $\rho A \Delta x$, where ρ is the density of the material of the string, and A its cross-sectional area. The acceleration in the vertical direction is

$$a = \frac{\partial^2}{\partial t^2} u.$$

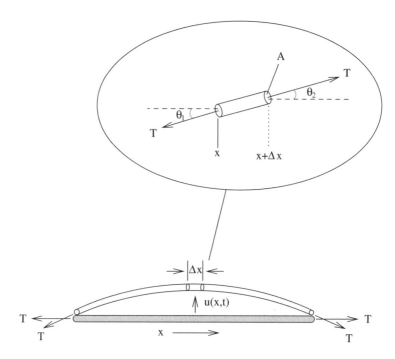

Figure 14.2. A stretched elastic string. (Drawing by E. Hinkle.)

The force should be the vertical component of the tension, plus other forces such as gravity and air friction.

The net vertical component of tension is

$$T \sin \theta_2 - T \sin \theta_1$$

$$\cong T[\theta_2 - \theta_1]$$

$$\cong T[u_x(x + \Delta x, t) - u_x(x, t)],$$

assuming that the angles θ_1 and θ_2 are small. The subscript x denotes partial derivative with respect to x. Putting these all together, we have

$$\rho A \Delta x \frac{\partial^2}{\partial t^2} u = TA[u_x(x + \Delta x, t) - u_x(x, t)] + \rho A \Delta x \cdot f, \qquad (14.1)$$

where f represents all additional force per unit mass. Equation (14.1) is

$$\frac{\partial^2}{\partial t^2} u = \frac{T}{\rho} \frac{1}{\Delta x}[u_x(x + \Delta x, t) - u_x(x, t)] + f,$$

which becomes, as $\Delta x \to 0$:

$$\frac{\partial^2}{\partial t^2} u = c^2 \frac{\partial^2}{\partial x^2} u + f, \qquad (14.2)$$

where $c^2 \equiv T/\rho$.

The tension along the bridge, T, is assumed to be uniform and is therefore equal to the force per unit area exerted on the suspension point $x = 0$ or $x = L$. Since the weight of the bridge is borne by these two suspension points, the vertical force exerted on each is half the weight of the bridge, and this should be equal to the projection of T in the vertical direction (see Figure 14.1):

$$T \sin \alpha = \frac{1}{2}(\rho L A)g/A = \frac{1}{2}\rho Lg,$$

where α is the angle from the horizontal to the tangent of the bridge at the suspension point, $g = 980 \,\text{cm/s}^2$, ρ is the density of the bridge material, and A is the cross section of the bridge. Thus

$$\boxed{c^2 \equiv T/\rho = \frac{1}{2}Lg/\sin\alpha.} \qquad (14.3)$$

Since the static weight of the bridge is balanced by the tension, the forcing f in (14.2) represents unbalanced vertical acceleration due to the marching soldiers. The system we need to solve is, with $u(x, t)$ being the vertical displacement of the bridge with respect to its equilibrium position:

$$\boxed{\begin{aligned} &\text{PDE:} \quad u_{tt} = c^2 u_{xx} + f(x, t), \quad 0 < x < L, \ t > 0, \\ &\text{BCs:} \quad u(0, t) = 0, \quad u(L, t) = 0, \quad t > 0, \\ &\text{ICs:} \quad u(x, 0) = 0, \quad u_t(x, 0) = 0, \quad 0 < x < L. \end{aligned}} \qquad (14.4)$$

The simplest expression for the periodic force exerted by a column of marching soldiers is probably

$$f(x, t) = a \sin(\omega_D t) \sin(\pi x/L), \quad 0 < x < L,$$

$$\omega_D = 2\pi/P. \qquad (14.5)$$

(Actually, this is meant to be the force *anomaly*, i.e., the difference between the force exerted by the marching soldiers and their static weight. This is why (14.5) can take on positive and negative values. The force due to the static weight of the soldiers, if it is a significant fraction of the weight of the bridge, can be incorporated into the weight of the bridge in our earlier calculation of the tension T. Nevertheless, the parameter c^2 in (14.3) should not be affected, amazingly!) Note that the assumed forcing implies that the soldiers march in synch. That is, the anomalous forcing is of the same sign across the span of the bridge.

Solution

There is a general method for solving boundary value problems called the eigenfunction expansion method. However, since knowledge of partial differential equations is not a prerequisite for this course, we shall proceed more intuitively.

The solution $u(x, t)$ is a function of both space and time. Given that the equation (14.4) is linear and contains a forcing term with a known x-structure:

$$X_1(x) \equiv \sin(\pi x/L), \quad 0 < x < L,$$

we shall try a solution of the form

$$u(x, t) = T_1(t)X_1(x). \tag{14.6}$$

Furthermore, (14.6) satisfies the boundary condition of u vanishing at $x = 0$ and $x = L$. We therefore do not need to worry about that boundary condition anymore. Substituting (14.6) into the partial differential equation in (14.4), we obtain

$$T_1''(t)X_1(x) = c^2 T_1(t)X_1''(x) + a \sin(\omega_D t)X_1(x). \tag{14.7}$$

The prime denotes differentiation with respect to the arguments. Thus

$$X_1''(x) = \frac{d^2}{dx^2}X_1(x) = -(\pi/L)^2 X_1(x).$$

Canceling out $X_1(x)$ in (14.7), we are left with

$$\frac{d^2}{dt^2}T_1(t) + \omega_1^2 T_1(t) = a \sin(\omega_D t). \tag{14.8}$$

The "natural frequency" ω_1 of the bridge is given by

$$\omega_1 = c\pi/L. \tag{14.9}$$

We see that the natural frequency would have been different if the forcing structure $X_1(x)$ had a shorter wavelength. Equation (14.8) is the ordinary differential equation for the forced oscillator described in some physics textbooks. Here we have given a physical derivation of how the natural frequency of the oscillator is determined; it is related to the spatial structure of the oscillation (14.6). This piece of information is not available if the bridge is modeled by an ordinary differential equation.

The solution to Eq. (14.8) consists of particular plus homogeneous solutions. The homogeneous solution is

$$A_1 \sin \omega_1 t + B_1 \cos \omega_1 t,$$

while the particular solution can be obtained by trying

$$D \sin(\omega_D t)$$

and finding $D = a/(-\omega_D^2 + \omega_1^2)$ upon substituting into (14.8):

$$T_1(t) = A_1 \sin \omega_1 t + B_1 \cos \omega_1 t + \frac{a \sin(\omega_D t)}{\omega_1^2 - \omega_D^2}$$

$$= \frac{T_1'(0)}{\omega_1} \sin \omega_1 t + T_1(0) \cos \omega_1 t + \frac{a}{\omega_1^2 - (\omega_D)^2} \left[\sin \omega_D t - \frac{\omega_D}{\omega_1} \sin \omega_1 t \right].$$

Applying the initial conditions from Eq. (14.4), we have $T_1(0) = 0$, $T_1'(0) = 0$. Finally, the solution is

$$u(x, t) = \frac{a}{\omega_1^2 - \omega_D^2} \left[\sin(\omega_D t) - \frac{\omega_D}{\omega_1} \sin \omega_1 t \right] \sin \frac{\pi x}{L}. \tag{14.10}$$

Resonance

The solution (14.10) involves the interference of a forced frequency ω_D with a fundamental frequency ω_1. When the two frequencies get close to each other, the numerator and the denominator of (14.10) both approach zero. Their ratio as $\omega_D \to \omega_1$ is obtained by l'Hôpital's rule to be (see Appendix A.2)

$$u(x, t) = a\left[\frac{-t\cos\omega_1 t}{2\omega_1} + \frac{\sin\omega_1 t}{2\omega_1^2}\right]\sin\left(\frac{\pi x}{L}\right). \qquad (14.11)$$

The oscillation grows in amplitude linearly in time, leading, presumably, to the collapse of the bridge.

The fundamental frequency ω_1 of the bridge is given by

$$\omega_1 = c\pi/L = \pi\sqrt{\frac{1}{2}g/(L\sin\alpha)}.$$

Thus the natural period P_1 is given by

$$P_1 \equiv 2\pi/\omega_1 = \sqrt{8L\sin\alpha/g},$$

which is about 1 second for a bridge 10 meters long, if the bridge deck is nearly horizontal, say $\alpha \sim 10°$:

$$P_1 = \sqrt{8L\sin\alpha/g} \sim 1.1 \text{ second}.$$

This is close to the probable forcing period P, and resonance is likely. Note that there is no need for an exact match of the two frequencies to get an enhanced response. Try to convince yourself that the oscillation is magnified when ω_1 is close to ω_D in the solution of Eq. (14.10).

A Different Forcing Function

Unlike ordinary differential equation models of resonance, which assume some *given* natural frequency of the system, the partial differential equation model discussed above determines the resonant frequency by the physical parameters of the bridge (via T/ρ) and by the x-shape of the forcing function $f(x, t)$. In the previous model, it was assumed that

$$f(x, t) = a\sin(\omega_D t)\sin(\pi x/L), \quad 0 < x < L.$$

So the forcing function has the shape as the first fundamental harmonic of the homogeneous system. Consequently, resonance occurs when the forcing frequency ω_D equals the frequency ω_1 of this fundamental mode. If we had instead used

$$f(x, t) = a\sin(\omega_D t)\sin(2\pi x/L), \quad 0 < x < L,$$

for our forcing function, resonance would have occurred when the forcing frequency ω_D equalled the frequency ω_2 of the second fundamental mode. (This is because we would have assumed

$$u(x, t) = T_2(t) X_2(t), \qquad X_2(x) \equiv \sin(2\pi x/L),$$

instead of (14.6) to match the x-structure of the forcing function. As a consequence, the natural frequency in (14.9) would have to be replaced by $\omega_2 = 2c\pi/L$.)

14.3 Tacoma Narrows Bridge

Even though numerous physics and mathematics textbooks attribute the 1940 collapse of the Tacoma Narrows Bridge to "a resonance between the natural frequency of oscillation of the bridge and the frequency of wind-generated vortices that pushed and pulled alternately on the bridge structure" (Halliday and Resnick, 1988), that bridge probably did *not* collapse for this reason (see Billah and Scanlan, 1991). As observed by Professor Burt Farquharson of the University of Washington, the wind speed at the time was 42 mph, giving a frequency of forcing by the vortex shedding mechanism of about 1 Hz. Professor Farquharson also observed that the frequency of the oscillation of the bridge just prior to its destruction was about 0.2 Hz. There was a mismatch of the two frequencies, and consequently this simple resonance mechanism probably was not the cause of the bridge's collapse. The bridge collapsed due to a torsional (twisting) vibration, as can be seen in old films and in Figure 14.3.

During its brief lifetime late in 1940, the bridge, under low-speed winds of 3–4 mph, did experience vertical modes of vibration that can probably be modeled by a model similar to the one presented here. However, the bridge endured this excited vibration *safely*. In fact, the bridge's nickname, "Galloping Gertie," was gained from such vertical motions under low wind, and this phenomenon occurred repeatedly since its opening day. Motorists crossing the bridge sometimes experienced a roller-coaster-like sensation as they watched cars ahead disappear from sight, then reappear. Tourists came from afar to experience it without worrying about their safety. Although the bridge had often "galloped," it had never twisted until November 7, 1940. At higher winds of 25–35 mph, there would be no oscillation of the bridge span. On November 7, 1940, under still heavier winds of 35–40 mph, the motion of the bridge turned into torsional oscillations. One sidewalk was raised 28 ft. above the other sidewalk. This lasted for about half an hour before the center span collapsed.

Figure 14.3. Twisting of Tacoma Narrows Bridge just prior to failure.

The above discussion points to the fact that although simple linear theories of forced resonance can perhaps explain the initial excitation of certain modes of oscillation, they cannot always be counted on to explain the final collapse of bridges, which is a very nonlinear phenomenon.

Assignment

Read McKenna (1999), which describes a nonlinear model of torsional oscillations.

14.4 Exercises

1.   Consider the problem of a column of soldiers marching across a suspension bridge of length L. The marching is slightly out of step, so the force exerted by the soldiers in the front of the column is opposite that in the rear. A simple model of the forcing term on the bridge is

$$f(x, t) = a \sin(2\pi t/P) \sin(2\pi x/L), \quad 0 < x < L.$$

Solve:

$$\text{PDE:} \quad u_{tt} = c^2 u_{xx} + f(x, t), \quad 0 < x < L, \quad t > 0,$$

$$\text{BCs:} \quad u(0, t) = 0, \quad u(L, t) = 0, \quad t > 0,$$

$$\text{ICs:} \quad u(x, 0) = 0, \quad u_t(x, 0) = 0, \quad 0 < x < L.$$

Discuss the criteria for resonance and sketch the shape of the mode excited.

2.   The discussion in the text points to the importance of modeling the forcing function realistically. A better model for $f(x, t)$ than (14.5) is probably

$$f(x, t) = a \sin(2\pi t/P), \quad 0 < x < L,$$

which assumes that the force exerted by the soldiers marching in unison is independent of where they are on the bridge. This seemingly simpler forcing function actually has a richer eigenfunction expansion. Let

$$f(x, t) = \sum_{n=1}^{\infty} f_n(t) \sin \frac{n\pi x}{L}, \quad 0 < x < L,$$

where (no need for you to show this)

$$f_n(t) = \begin{cases} 0 & \text{if } n \text{ is even,} \\ \dfrac{4a}{n\pi} \sin(2\pi t/P) & \text{if } n \text{ is odd.} \end{cases}$$

a. Verify that the solution to Eq. (14.4) now becomes

$$u(x, t) = \sum_{n=1}^{\infty} T_n(t) \sin \frac{n\pi x}{L},$$

where

$$T_n(t) = 0$$

if n is even, and

$$T_n(t) = \frac{4a/\pi}{\omega_n^2 - \omega_D^2}\left[\sin(\omega_D t) - \frac{\omega_D}{\omega_n}\sin \omega_n t\right]/n$$

if n is odd.

b. There are now chances for resonance whenever

$$P = 2\pi/\omega_n \text{ for some } n.$$

However, because the amplitude of $T_n(t)$ decreases with n, probably only the first two modes will have any real impact. To resonate the first harmonic mode, what must the forcing period P be?

c. The next nonzero fundamental mode is the third one. To resonate with this mode, what must the forcing P be?

d. A column of soldiers running in unison with a third of a second between steps may be able to induce an oscillation in the third mode. Can this mode possibly be near resonance? What would be the x shape of the resulting oscillation of the bridge? How does it compare with the x-shape of the forcing?

3.   Let $N(x, t)$ be the mass of red-tide algae per unit mass of sea water. In the absence of transport by wind waves, the algae grow according to

$$\frac{\partial}{\partial t}N = r N,$$

where r is the biological production rate. In the presence of diffusive transport, which we are now considering, the growth is different at different locations. It is now governed by the following system:

PDE: $\dfrac{\partial}{\partial t}N = r N + D\dfrac{\partial^2}{\partial x^2}N, \quad 0 < x < L,$

BC: $N(0, t) = 0, \quad N(L, t) = 0.$

The quantity D is the coefficient of diffusion by the random wind waves. The boundary condition is meant to simulate the fact that favorable conditions for algae growth exist only in the strip $0 < x < L$. The algae will be killed quickly if they are transported beyond this strip of ocean.

a. Substitute $N(x, t) = X(x)T(t)$ into the partial differential equation and the boundary condition to find $X(x)$ and $T(t)$. Note that there are n such pairs. The general solution is a superposition of all $X_n(x)T_n(t)$.

b. Show that the algae will become extinct if $L < \pi \sqrt{D/r}$ and that there is an explosive outbreak of red tide otherwise.

APPENDIX A

Differential Equations and Their Solutions

A.1 First- and Second-Order Equations

Example: Population Growth

$$\frac{dN}{dt} = r N, \tag{A.1}$$

where $N(t)$ is the population density of a species at time t. The above equation is simply a statement that the rate of population growth, $\frac{dN}{dt}$, is proportional to the population itself, with the proportionality constant r. To solve it, we move all the N's to one side and all the t's to the other side of the equation. (This process is called "separation of variables" in the ordinary differential equation literature. We will not use this term here, as it may get confused with a partial differential equation solution method with the same name, which we will discuss later.) Thus

$$\frac{dN}{N} = r\,dt. \tag{A.2}$$

Integrating both sides yields

$$\ln N = rt + \text{constant},$$

which can be rewritten as

$$N(t) = \text{constant} \cdot e^{rt}.$$

In order for the left-hand side to equal the right-hand side at $t = 0$, the "constant" in the second equation must be $N(0)$. Thus,

$$N(t) = N(0)e^{rt}. \tag{A.3}$$

The population grows exponentially from an initial value $N(0)$, with an e-folding time of r^{-1}. That is, $N(t)$ increases by a factor of e with every increment of r^{-1} in t. The solution is plotted in Figure A.1.

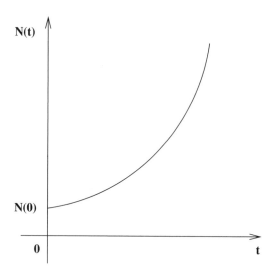

Figure A.1. Solution to Eq. (A.1).

Equation (A.1) is perhaps an unrealistic model for most population growth. Among other things, its solution implies that the population will grow indefinitely. A better model is given by the following equation:

$$\frac{dN}{dt} = r N \cdot (1 - N/k).$$ (A.4)

Try solving it using the same method. The solution is plotted in Figure A.2 for $0 < N(0) < k$.

Example: Harmonic Oscillator Equation

$$m\frac{d^2x}{dt^2} + kx = 0.$$ (A.5)

Here x is the vertical displacement from the equilibrium position of the spring, which has a mass m and a spring constant k. Equation (A.5) is a statement of Newton's law of motion: $m\frac{d^2x}{dt^2}$ is mass times acceleration. This is required to be equal to the spring's restoring force $-kx$. This force is assumed to be proportional to the displacement from equilibrium.

It is harder to solve a second-order ordinary differential equation, although this particular equation is so common that we were taught

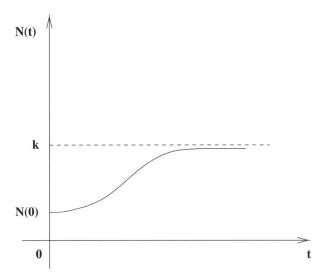

Figure A.2. Solution to Eq. (A.4).

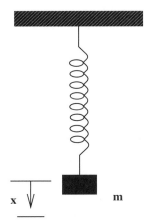

Figure A.3. A spring of mass m suspended under gravity and in equilibrium. x is the displacement from the equilibrium position.

to try exponential solutions whenever we see linear equations with constant coefficients.

Solution by trial method: Try $e^{\alpha t}$

Plugging the guess into Eq. (A.5) for x then suggests that α must satisfy

$$m\alpha^2 + k = 0,$$

which means

$$\alpha = i\sqrt{k/m} \quad \text{or} \quad \alpha = -i\sqrt{k/m},$$

where $i \equiv \sqrt{-1}$ is the imaginary number. There are two different values of α that will make our trial solution satisfy Eq. (A.5). So the general solution should be a linear combination of the two:

$$x(t) = c_1 e^{i\sqrt{k/m}t} + c_2 e^{-i\sqrt{k/m}t}, \qquad (A.6)$$

where c_1 and c_2 are two arbitrary (complex) constants.

If you do not like using complex notation (numbers that involve i), you can rewrite (A.6) in real notation, making use of Euler's identity, which we will derive a little later:

$$e^{i\theta} = \cos\theta + i\sin\theta. \qquad (A.7)$$

Thus the previous solution (A.6) can be rewritten as

$$x(t) = A\sin(\sqrt{k/m}t) + B\cos(\sqrt{k/m}t), \qquad (A.8)$$

where A and B are some arbitrary real constants (since c_1 and c_2 were undetermined).

We can verify that (A.8) is indeed the solution to the harmonic oscillator equation (A.5) by noting, from calculus,

$$\frac{d}{dt}\sin\omega t = \omega\cos\omega t, \quad \frac{d}{dt}\cos\omega t = -\omega\sin\omega t,$$

so

$$\frac{d^2}{dt^2}\sin\omega t = \frac{d}{dt}(\omega\cos\omega t) = -\omega^2\sin\omega t$$

and

$$\frac{d^2}{dt^2}\cos\omega t = \frac{d}{dt}(-\omega\sin\omega t) = -\omega^2\cos\omega t.$$

Therefore, the sum (A.8) satisfies

$$\frac{d^2}{dt^2}x = -\omega^2 x, \tag{A.9}$$

which is the same as (A.5), provided that $\omega^2 = k/m$.

Euler's Identity

Euler's Identity, as used in (A.7), deserves some comment. In calculus, we learned how to differentiate an exponential

$$\frac{d}{d\theta}e^{a\theta} = ae^{a\theta}.$$

Although you have always assumed a to be a real number, it does not make any difference if a is complex. So, letting $a = i$, we find

$$\frac{d}{d\theta}e^{i\theta} = ie^{i\theta},$$

$$\frac{d^2}{d\theta^2}e^{i\theta} = \frac{d}{d\theta}(ie^{i\theta}) = i^2e^{i\theta} = -e^{i\theta}.$$

We have thus shown that the function

$$y(\theta) = e^{i\theta} \tag{A.10}$$

satisfies the second-order ordinary differential equation:

$$\frac{d^2}{d\theta^2}y + y = 0. \tag{A.11}$$

$e^{i\theta}$ also happens to satisfy the *initial conditions*:

$$y(0) = 1, \quad \frac{d}{d\theta}y(0) = i. \tag{A.12}$$

On the other hand, we have just verified in (A.9) that

$$y = A\sin\theta + B\cos\theta \tag{A.13}$$

also satisfies (A.11), which is the same as (A.9) if we replace t by θ and ω by 1. If we furthermore require the sum (A.13) to also satisfy the initial condition (A.12), we will find that $B = 1$ and $A = i$. Since

$$y(\theta) = \cos\theta + i\sin\theta \tag{A.14}$$

satisfies the same ordinary differential equation (A.11) and the same initial conditions (A.12) as (A.10), (A.14) and (A.10) must be the same by the uniqueness theorem for ordinary differential equations. What we have outlined is one way for proving the Euler identity (A.7):

$$e^{i\theta} = \cos\theta + i\sin\theta.$$

Solution from First Principles

If you prefer to find the solution from first principles, i.e., not by guessing that it should be in the form of an exponential, it can be done by "reduction of order," although we normally do not bother to do it this way.

We recognize that Eq. (A.5) is a type of ordinary differential equation with its "independent variable missing." The method of reduction of order suggests that we let

$$p \equiv \frac{dx}{dt},$$

and write

$$\frac{d^2x}{dt^2} = \frac{dp}{dt} = \frac{dx}{dt}\frac{dp}{dx} = p\frac{dp}{dx}.$$

Treating p now as a function of x, Eq. (A.5) becomes

$$p\frac{dp}{dx} + \omega^2 x = 0, \qquad (A.15)$$

where we have used ω^2 for k/m.

Equation (A.15), a first-order ordinary differential equation, can be solved by the same method we used in Example 1, as follows.

Integrating $pdp + \omega^2 xdx = 0$ yields

$$p^2 + \omega^2 x^2 = \omega^2 a^2,$$

with $\omega^2 a^2$ being an arbitrary constant of integration. From $p = \pm\omega\sqrt{a^2 - x^2}$, we have, since $p = \frac{d}{dt}x$,

$$\frac{dx}{dt} = \pm\omega\sqrt{a^2 - x^2}.$$

This is again a first-order ordinary differential equation, which we solve as before.

Integrating both sides of

$$\frac{dx}{\sqrt{a^2 - x^2}} = \pm\omega dt$$

and using the integral formula:

$$\int \frac{dx}{\sqrt{a^2 - x^2}} = \sin^{-1}\frac{x}{a} + b, \quad b \text{ being a constant,}$$

we find

$$\sin^{-1}\frac{x}{a} = \pm\omega t + b.$$

Inverting, we find

$$\frac{x}{a} = \sin(\pm\omega t + b),$$

which can (finally!) be rewritten as

$$x(t) = A\sin\omega t + B\cos\omega t. \tag{A.16}$$

A.2 Nonhomogeneous Ordinary Differential Equations

First-Order Equations

A nonhomogeneous version of the example in (A.1) is

$$\boxed{\frac{dN}{dt} - rN = f(t),} \tag{A.17}$$

where $f(t)$ is a (known) specified function of t, independent of the "unknown" N. It is called the "inhomogeneous term" or the "forcing term." In the population growth example we discussed earlier, $f(t)$ can represent the rate of population growth of the species due to migration.

We are here concerned with the method of solution of (A.17) for any given $f(t)$. We proceed to multiply both sides of (A.17) by a yet-to-be-determined function $\mu(t)$, called the *integrating factor*:

$$\mu\frac{dN}{dt} - r\mu N = \mu f. \tag{A.18}$$

We choose $\mu(t)$ such that the product on the left-hand side of (A.18) is a perfect derivative, i.e.,

$$\mu \frac{dN}{dt} - r\mu N = \frac{d}{dt}(\mu N). \tag{A.19}$$

If this can be done, then (A.10) would become

$$\frac{d}{dt}(\mu N) = \mu f,$$

which can be integrated from $t = 0$ to t to yield

$$\mu(t)N(t) - \mu(0)N(0) = \int_0^t \mu(t)f(t)\,dt. \tag{A.20}$$

The notation on the right-hand side of (A.20) is rather confusing. A better way is to use a different symbol, say τ, in place of t as the dummy variable of integration. Then, (A.20) can be rewritten as

$$N(t) = N(0)\mu(0)\mu^{-1}(t) + \mu^{-1}(t)\int_0^t \mu(\tau)f(\tau)\,d\tau. \tag{A.21}$$

The remaining task is to find the integrating factor $\mu(t)$. In order for (A.19) to hold, we must have the right-hand side

$$\mu \frac{dN}{dt} + N\frac{d\mu}{dt}$$

equal the left-hand side, implying

$$\frac{d\mu}{dt} = -r\mu. \tag{A.22}$$

The solution to (A.22) is simply

$$\mu(t) = \mu(0)e^{-rt}. \tag{A.23}$$

Substituting (A.23) back into (A.21) then yields

$$\boxed{N(t) = N(0)e^{rt} + e^{rt}\int_0^t e^{-r\tau} f(\tau)\,d\tau.} \tag{A.24}$$

This then completes the solution of (A.17). If r is not a constant but is a function t, the procedure remains the same up to and including (A.22). The solution to (A.22) should now be

$$\mu(t) = \mu(0)e^{-\int_0^t r(t')dt'}. \tag{A.25}$$

The final solution is obtained by substituting (A.25) into (A.21).

Notice that the solution to the (linear) nonhomogeneous equation consists of two parts: a part satisfying the general homogeneous equation and a part that is a particular solution of the nonhomogeneous equation (the first and second terms on the right-hand side of (A.24), respectively). For some simple forcing functions $f(t)$, there is no need to use the general procedure of integrating factors if we can somehow guess a particular solution. For example, suppose we want to solve

$$\boxed{\frac{dN}{dt} - rN = 1.} \tag{A.26}$$

We write

$$N(t) = N_h(t) + N_p(t),$$

where $N_h(t)$ satisfies the homogeneous equation

$$\frac{dN_h}{dt} - rN_h = 0$$

and so is

$$N_h(t) = ke^{rt}$$

for some constant k. $N_p(t)$ is any solution of Eq. (A.26). By the "method of judicious guessing," we try selecting a constant for $N_p(t)$. Upon substituting $N_p(t) = a$ into (A.26), we find the only possibility: $a = -\frac{1}{r}$. Thus the full solution is

$$N(t) = ke^{rt} - \frac{1}{r} = N(0)e^{rt} + \frac{1}{r}(e^{rt} - 1).$$

Second-Order Equations

For our purpose here, we will be using only the "method of judicious guessing" for linear second-order equations. The more general method of "variation of parameters" is too cumbersome for our limited purposes.

Example 1
Solve

$$\frac{d^2}{dt^2}x + \omega^2 x = 1. \tag{A.27}$$

We write the solution as a sum of a homogeneous solution x_h and a particular solution x_p, i.e.,

$$x(t) = x_h(t) + x_p(t).$$

We already know that the homogeneous solution (to (A.5)) is of the form

$$x_h(t) = A\sin\omega t + B\cos\omega t. \tag{A.28}$$

We guess that a particular solution to (A.27) is a constant,

$$x_p(t) = a. \tag{A.29}$$

Subsituting (A.29) into (A.27) then shows that the constant is $1/\omega^2$. Thus the general solution to (A.27) is

$$\boxed{x(t) = A\sin\omega t + B\cos\omega t + 1/\omega^2.} \tag{A.30}$$

The arbitrary constants A and B are to be determined by initial conditions.

Example 2
Solve

$$\boxed{\frac{d^2}{dt^2}x + \omega^2 x = \sin\omega_0 t.} \tag{A.31}$$

Again we write the solution as a sum of the homogeneous solution and a particular solution. For the particular solution, we try

$$x_p(t) = a \sin \omega_0 t. \tag{A.32}$$

Upon substitution of (A.32) into (A.31), we find

$$a = (\omega^2 - \omega_0^2)^{-1},$$

and so the general solution is

$$\boxed{x(t) = A \sin \omega t + B \cos \omega t + \frac{\sin \omega_0 t}{(\omega^2 - \omega_0^2)}.} \tag{A.33}$$

The solution in (A.33) is valid as it is for $\omega_0 \neq \omega$. Some special treatment is helpful when the forcing frequency ω_0 approaches the natural frequency ω. Let us write

$$\omega_0 = \omega + \epsilon$$

and let $\epsilon \to 0$. The particular solution can be written as

$$x_p(t) = \frac{\sin \omega_0 t}{(\omega^2 - \omega_0^2)} = \frac{\sin(\omega t + \epsilon t)}{\omega^2 - (\omega + \epsilon)^2}$$

$$= \frac{\sin \omega t \cos \epsilon t + \cos \omega t \sin \epsilon t}{-2\omega\epsilon - \epsilon^2} \to -\frac{\cos \omega t}{2\omega} \cdot t - \frac{\sin \omega t}{2\omega\epsilon} \quad \text{as } \epsilon \to 0.$$

Thus, for the case of *resonance*, $\omega_0 = \omega$, the solution (1.33) becomes

$$x(t) = A' \sin \omega t + B \cos \omega t - \frac{1}{2\omega} t \cos \omega t,$$

where we have written $A = A' + \frac{1}{2\omega\epsilon}$, with A' being a (finite) arbitrary constant. The solution grows secularly in t.

A.3 Summary of ODE Solutions

In this book we will be mostly dealing with linear differential equations with constant coefficients. For these, simply try an exponential solution. This is the easiest way. You are not expected to have to repeat each time the derivation given in the previous sections on why the exponentials are the right solutions to try. Just do the following:

a. $\boxed{\frac{d}{dt}N = rN.}$

Try $N(t) = ae^{\alpha t}$ and find $\alpha = r$, so the solution is

$$\boxed{N(t) = ae^{rt}.}$$

b. $\boxed{\frac{d^2}{dt^2}x + \omega^2 x = 0.}$

Try $x(t) = ae^{\alpha t}$ and find $\alpha = \pm i\omega$ so the complex solution is

$$\boxed{x(t) = a_1 e^{i\omega t} + a_2 e^{-i\omega t}}$$

and the real solution

$$x = A\cos\omega t + B\sin\omega t.$$

A.4 Exercises

1. Find the most general solution to the following conditions.

    a. $\dfrac{d}{dt}N = rN + b$; r, b are constants;

    b. $\dfrac{d^2}{dt^2}x + \beta\dfrac{d}{dt}x + \omega^2 x = 0$; β, ω^2 are constants;

    c. $\dfrac{d^2}{dt^2}x + \omega^2 x = \cos\omega_0 t$, $\omega \neq \omega_0$;

    d. $\dfrac{d^2}{dt^2}x + \omega_0^2 x = \cos\omega_0 t$.

2.   Find the solution satisfying the specified initial conditions:

a. $\frac{d}{dt}N = rN + b$,

N(0) = 0.

b. $\frac{d^2}{dt^2}x + \beta\frac{d}{dt}x + \omega^2 x = 0$,

$x(0) = 1, \frac{d}{dt}x(0) = 0$.

c. $\frac{d^2}{dt^2}x + \omega^2 x = \cos\omega_0 t, \omega \neq \omega_0$,

$x(0) = 0, \frac{d}{dt}x(0) = 0$.

d. $\frac{d^2}{dt^2}x + \omega_0^2 x = \cos\omega_0 t$,

$x(0) = 0, \frac{d}{dt}x(0) = 0$.

A.5 Solutions to Exercises

1.a. $N(t) = N_h(t) + N_p(t)$, where $N_h(t)$ is the solution to the homogeneous equation and $N_p(t)$ is a particular solution to the nonhomogeneous equation. For $N_p(t)$, we try a constant, i.e., $N_p(t) = c$. Substituting it into $\frac{d}{dt}N = rN + b$ yields $0 = rc + b$, implying $c = -b/r$.

The solution to the homogeneous equation $\frac{d}{dt}N = rN$ is $N_h(t) = ae^{rt}$, where a is an arbitrary constant. Combining, we get

$$N(t) = ae^{rt} - b/r.$$

b. Try

$$x(t) = ae^{\alpha t}.$$

Substituting into the ordinary differential equation yields

$$\alpha^2 + \alpha\beta + \omega^2 = 0.$$

So $\alpha = \alpha_1$ or $\alpha = \alpha_2$, where

$$\alpha_1 \equiv -\frac{\beta}{2} + \sqrt{\frac{\beta^2}{4} - \omega^2} \quad \text{and} \quad \alpha_2 \equiv -\frac{\beta}{2} - \sqrt{\frac{\beta^2}{4} - \omega^2}.$$

The general solution is

$$x(t) = a_1 e^{\alpha_1 t} + a_2 e^{\alpha_2 t}.$$

c. $x(t) = x_h(t) + x_p(t)$.
For $x_p(t)$, try $x_p(t) = a \cos \omega_0 t$. Substituting into the ordinary differential equation yields

$$-\omega_0^2 a + \omega^2 a = 1.$$

So $a = 1/(\omega^2 - \omega_0^2)$. For $x_h(t)$, we know the general solution to the homogeneous ordinary differential equation is

$$x_h(t) = A \sin \omega t + B \cos \omega t; \; A, B \text{ are arbitrary constants.}$$

The full solution is

$$x(t) = A \sin \omega t + B \cos \omega t + \cos \omega_0 t / (\omega^2 - \omega_0^2).$$

d. This is the resonance case. Still try $x(t) = x_h(t) + x_p(t)$.
For $x_p(t)$, try $x_p(t) = at \sin \omega_0 t$. Substituting into the nonhomogeneous ordinary differential equation yields

$$2a\omega_0 \cos \omega_0 t - a\omega_0^2 t \sin \omega_0 t + \omega_0^2 at \sin \omega_0 t = \cos \omega_0 t.$$

Thus

$$2a\omega_0 = 1.$$

So $x_p(t) = t \sin \omega_0 t / 2\omega_0$. The general solution to the homogeneous ordinary differential equation is

$$x_h(t) = A \sin \omega_0 t + B \cos \omega_0 t.$$

The full solution is

$$x = A \sin \omega_0 t + B \cos \omega_0 t + t \sin \omega_0 t / 2\omega_0.$$

2.a. $N(t) = ae^{rt} - b/r$.

$$N(0) = a - b/r = 0 \text{ implies } a = b/r.$$

So,

$$N(t) = \frac{b}{r}(e^{rt} - 1).$$

b. $x(t) = a_1 e^{\alpha_1 t} + a_2 e^{\alpha_2 t}$.

$$x(0) = a_1 + a_2 = 1,$$

$$\frac{d}{dt}x(0) = \alpha_1 a_1 + \alpha_2 a_2 = 0.$$

Thus $a_1 = -\alpha_2/(\alpha_1 - \alpha_2)$, $a_2 = \alpha_1/(\alpha_1 - \alpha_2)$. Finally

$$x(t) = \frac{1}{(\alpha_1 - \alpha_2)}[-\alpha_2 e^{\alpha_1 t} + \alpha_1 e^{\alpha_2 t}].$$

c. $x(t) = A \sin \omega t + B \cos \omega t + \cos \omega_0 t/(\omega^2 - \omega_0^2)$.

$$x(0) = B + 1/(\omega^2 - \omega_0^2),$$

$$\frac{d}{dt}x(0) = A\omega = 0.$$

Thus $A = 0$, $B = -1/(\omega^2 - \omega_0^2)$, and

$$x(t) = \frac{1}{(\omega^2 - \omega_0^2)}[\cos \omega_0 t - \cos \omega t].$$

d. $x(t) = A \sin \omega_0 t + B \cos \omega_0 t + t \sin \omega_0 t/2\omega_0$.

$$x(0) = B = 0,$$

$$\frac{d}{dt}x(0) = A\omega_0 = 0.$$

Thus $A = 0$, $B = 0$, and

$$x(t) = t \sin \omega_0 t/2\omega_0.$$

APPENDIX B

MATLAB Codes

B.1 MATLAB Codes for Lorenz Equations

```
%%%%%%%%%%%%%%%%%%%%%%%%%%%%%%%%%%%%%%%%%%%%%%%%%%%%%%%
% solves the Lorenz equations with ode45            %
% requires the use of file:  myode.m                %
% produces two figures:                             %
% figure 1 = x,y,z vs. time                         %
% figure 2 = 3D trajectory of "lorenz butterfly"    %
%%%%%%%%%%%%%%%%%%%%%%%%%%%%%%%%%%%%%%%%%%%%%%%%%%%%%%%

clear;
% declare global parameters
global r;
global sigma;
global b;

% set parameters
r = 28;
sigma = 10;
b = 8/3;

% initial conditions
x0 = 3;
y0 = 15;
z0 = 1;
v0 = [x0,y0,z0];
t0 = 0;
tf = 60;
tspan = [t0 tf];
%solves 'myode' using ode45 routine

[t,vout] = ode45('myode',tspan,v0);

% note:
paramstring = sprintf( ', {\\itr}=%d, {\\it\\sigma}=%d,
{\\itb}=%2.2f', r, sigma, b );
icstring = sprintf( ', ({\\itx}_0,{\\ity}_0,
```

```
{\\itz}_0)=(%d,%d,%d)', x0, y0, z0 );
figure(1)
clf
%plots x(t) in the top 1/3 of the figure
subplot(3,1,1)

plot( t, vout(:,1) );
xlabel( '{\itt}' );
ylabel( '{\itx}({\itt})' );
title( ['Lorenz Solutions', paramstring, icstring ] )
axis tight;
grid on;
%plots y (t) in the second 1/3 of the figure

subplot(3,1,2);
plot( t, vout(:,2) );
xlabel( '{\itt}' );
ylabel( '{\ity}({\itt})' );
axis tight;
grid on;
%plots z (t) in the third 1/3 of the figure

subplot(3,1,3);
plot( t, vout(:,3) );
xlabel( '{\itt}' );
ylabel( '{\itz}({\itt})' );
axis tight;
grid on;

figure(2);
clf;
% plots the trajectory
plot3( vout(:,1), vout(:,2), vout(:,3) );
hold on;
% plot a circle at initial point
plot3( vout(1,1), vout(1,2), vout(1,3), 'o' );

axis tight;
xlabel( '{\itx}' );
ylabel( '{\ity}' );
zlabel( '{\itz}' );
title( ['Lorenz Trajectory', paramstring, icstring] );
```

```
camproj('perspective');
campos([180.8444 -309.0346  241.0763] );
camtarget( [1.6213    3.2901    25.8520] );

function vprime = myode( t, v )
% parameters (defined elsewhere)
global sigma;
global r;
global b;

% map current vector to user-friendly variable names
x = v(1);
y = v(2);
z = v(3);

% compute derivatives with the Lorenz model
xprime = sigma*(-x+y);
yprime = r*x-y-x*z;
zprime = -b*z + x*y;

% map back to vector format
vprime = zeros(3,1);
vprime(1) = xprime;
vprime(2) = yprime;
vprime(3) = zprime;
```

B.2 MATLAB Codes for Solving Vallis's Equations

```
%%%%%%%%%%%%%%%%%%%%%%%%%%%%%%%%%%%%%%%%%%%%%%%%%%%%%%%%%%%%%%%%%
%
%     solves the set of equations
%
%        x' = b y - c (x - xstar)
%        y' = x z - y
%        z' = -x y - (z - 1)
%
%     The form of the equations is defined by the function
%     'Vallisode'
%
%%%%%%%%%%%%%%%%%%%%%%%%%%%%%%%%%%%%%%%%%%%%%%%%%%%%%%%%%%%%%%%%%
```

```
      clear

      global b;
      global c;
      global xstar;

      b = 102;
      c = 3;
      xstar = 0;
% time span of integration
      tspan = [0 60];
%     set initial conditions
      x0=0
      y0=2.0/12.0
      z0=1.0

      v0  = [x0 y0 z0];    % initial conditions for x,y,z

      [t,vout] = ode45('Vallisode',tspan,v0);

%
%     plots y(t) for xstar=0.0 in an upper subplot
%

% note:
      paramstring = sprintf( ', {\\itb}=%d, {\\itc}=%d,
      {\\itx_0}=%d', b, c, xstar );

      figure(1)
      clf
      subplot(2,1,1)

      plot( t, vout(:,2) );
      xlabel( '{\itt}' );
      ylabel( '{\itx}({\itt})' );
      title( ['East-west Temperature Difference', paramstring ] )
      axis tight;
      grid on;
%%%%%%%%%%
      xstar = -0.83;

      [t,vout] = ode45('Vallisode',tspan,v0);
```

```
%
%       plots y(t) for specified xstar in a lower subplot
%
%

        paramstring = sprintf( ', {\\itb}=%d, {\\itc}=%d,
        {\\itx_0}=%2.2f', b, c, xstar );
        subplot(2,1,2);
        plot( t, vout(:,2) );
        xlabel( '{\itt}' );
        ylabel( '{\ity}({\itt})' );
        title( ['East-west Temperature Difference', paramstring ] )
        axis tight;
        grid on;
%%%%%%%%%%%%%%%%%%%%%


%
        function  vprime = Vallisode(t,vin)

        global b;
        global c;
        global xstar;

        x = vin(1);
        y = vin(2);
        z = vin(3);

        xprime = b*y - c*(x-xstar);
        yprime = x*z - y;
        zprime = -x*y - (z-1);

        vprime = [xprime; yprime; zprime];
```

Bibliography

Adamson, J., 1961: *Living Free: The Story of Elsa and Her Cubs*, Collins and Harvill (London), p 29.

Atela, P., C. Golé, and S. Hotton, 2002: A dynamical system for plan pattern formation: Rigorous analysis, *J. Nonlinear Sci.*, **12**, (6), 1–39.

Banks, R., 1993: *Growth and Diffusion Phenomena: Mathematical Frameworks and Applications*, Springer-Verlag (Berlin).

Barabási, A., and R. Albert, 1999: Emergence of scaling in random networks, *Science*, **286**, 509–512.

Billah, K. Y., and R. H. Scanlan, 1991: Resonance, Tacoma-Narrows bridge failure, and undergraduate physics text, *Am. J. Phys.*, **59**, (2), 118–124.

Brown, J. H., G. B. West, and B. J. Enquist, 2005: Yes, West, Brown and Enquist's model of allometric scaling is both mathematically correct and biologically relevant, *Funct. Ecol.*, **19**, 735–738.

Budyko, M. I., 1969: The effect of solar radiation variation on the climate of the Earth, *Tellus*, **21**, 611–619.

Budyko, M. I., 1972: The future climate, *Trans. Am. Geophys. Union*, **53**, 868–874.

Burghes, D. N., I. Huntley, and J. McDonald, 1982: *Applying Mathematics: A Course in Mathematical Modelling*, Halsted Press (New York).

Cahalan, R. F., and G. R. North, 1979: A stability theorem for energy-balance climate models, *J. Atmos. Sci.*, **36**, 1178–1186.

Caldeira, K., and J. F. Kastings, 1992: Susceptibility of the early Earth to irreversible glaciation caused by carbon dioxide clouds, *Nature*, **359**, 226–228.

Cane, M. A., and S. E. Zebiak, 1985: A theory for El Niño and the Southern Oscillation, *Science*, **228**, 1084–1087.

Cane, M. A., S. E. Zebiak, and S. C. Dolan, 1986: Experimental forecasts of El Niño, *Nature*, **321**, 827–832.

Cayrel, R., V. Hill, T. C. Beers, B. Barbuy, M. Spite, F. Spite, B. Plez, J. Andersen, P. Bonifacio, P. François, P. Molaro, B. Nordström, F. Primas, 2001: Measurement of stellar age from uranimum decay, *Nature*, **409**, 691–692.

Chylek, P., and J. A. Coakley, 1975: Analytical analysis of a Budyko-type climate model, *J. Atmos. Sci.*, **32**, 675–679.

Cook, J., R. Tyson, J. White, R. Rushe, J. M. Gottman, and J. D. Murray, 1995: The mathematics of marital conflict: Qualitative dynamics mathematical modeling of marital interaction, *J. Family Psychol.*, **9**, 110–130.

Curry, J. H., J. R. Herring, J. Loncaric, and S. A. Orszag, 1984: Order and disorder in two- and three-dimensional Bénard convection, *J. Fluid Mech.*, **147**, 1–38.

Etienne, R. S., M. E. Apol, and H. Olff, 2006: Demystifying the West, Brown & Enquist model of the allometry of metabolism, *Funct. Ecol.*, **20**, 394–399.

Farmelo, G., ed., 2002: *It Must Be Beautiful: Great Equations of Modern Science*, Granta Books (London).

Farrell, B. F., 1990: Equable climate dynamics, *J. Atmos. Sci.*, **47**, 2986–2995.

Fischler, R., 1979: What did Herodotus say? Or how to build (a theory of) the great pyramid, *Environ. Planning*, **B6**, 89–93.

Flannery, T., 1993: The case of the missing meat eaters, *Natural History*, June, 22–24.

Frederiksen, J., 1976: Nonlinear albedo-temperature coupling in climate models, *J. Atmos. Sci.*, **33**, 2267–2272.

Gazale, M. J., 1999: *Gnomon: From Pharaohs to Fractals*, Princeton University Press (Princeton, NJ).

Gottman, J. M., 1979: *Marital Interaction: Empirical Investigations*, Academic Press (New York).

Gottman, J. M., 1994: *Why Marriages Succeed or Fail*, Simon and Schuster (New York).

Gottman, J. M., J. D. Murray, C. C. Swanson, R. Tyson, and K. R. Swanson, 2002: *The Mathematics of Marriage*, MIT Press (Cambridge, MA).

Graves, C. E., W.-H. Lee, and G. R. North, 1993: New parameterization and sensitivity for simple climate models, *J. Geophys. Res.*, **98**, 5025–5036.

Haberman, R., 1988: *Mathematical Models: Mechanical Vibrations, Population Dynamics, and Traffic Flow*, SIAM Classics, SIAM (Philadelphia).

Halliday, D., and R. Resnick, 1988: *Fundamentals of Physics*, Wiley (New York), Third Edition.

Hansen, J., et al., 1985: Climate response times: Dependence on climate sensitivity and ocean mixing, *Science*, **229**, 857–859.

Hansen, J., et al., 2005: Earth's energy imbalance: Confirmation and implications, *Science*, **308**, 1431–1435.

Harland, W. B., and M. J. S. Rudwick, 1964: The Great Infra-Cambrian Glaciation, *Scientific American*, **211**, (2), 28–36.

Hartmann, D. L., 1994: *Global Physical Climatology*, Academic Press (Boston).

Held, I. M., and M. J. Suarez, 1974: Simple albedo feedback models of icecaps, *Tellus*, **26**, 613–629.

Hoffman, P. F., A. J. Kauffman, G. P. Halverson, and D. P. Schrag, 1998: A Neoproterozoic snowball Earth, *Science*, **281**, 1342–1346.

Hoffman, P. F., and D. P Schrag, 2000: Snowball Earth, *Sci. Am.*, January, 68–75.

Hyde, W. T., T. J. Crowley, S. K. Baum, and R. Peltier, 2000: Neoproterozoic "snowball Earth" simulations with a coupled climate/ice-sheet model, *Nature*, **405**, 425–429.

IPCC (Intergovernmental Panel on Climate Change), 1990: *Climate Change, The IPCC Scientific Assessment*, J.T. Houghton et al., eds., Cambridge University Press (New York).

IPCC (Intergovernmental Panel on Climate Change), 2001: *Climate Change 2001: The Scientific Basis*, J. T. Houghton et al., eds., Cambridge University Press (New York).

Kirschvink, J. L., 1992: Late Proterozoic low-latitude global glaciation: The snowball Earth, in *The Proterozoic Biosphere*, J. W. Schopt and C. Klein, eds., Cambridge University Press (Cambridge, UK), 51–52.

Kot, M., 2001, *Elements of Mathematical Ecology*, Cambridge University Press (Cambridge, UK).

Kot, M., C. Berg, and E. Silverman, 2003: Zipf's law and the diversity of biology newsgroups, *Scientometrics*, **56**, 247–257.

Kozlowski, J., and M. Konarzewski, 2004: Is West, Brown and Enquist's model of allometric scaling mathematically correct and biologically relevant? *Funct. Ecol.*, **18**, 283–289.

Lanchester, F. W., 1914: *Mathematics in Warfare*, originally published in 1914; reprinted in *The World of Mathematics*, J. R. Newmann, ed., Simon and Schuster (New York, 1956), 2136–2157.

Laurent, P., 1917: The supposed synchronal flashing of fireflies, *Science*, **45**, 44.

Li, W., 1992: Random texts exhibit Zipf's law-like word frequency distribution, *IEEE Trans. Inf. Theory*, **38**, 1842–1845.

Lin, C. C., and L. A. Segal, 1974: *Mathematics Applied to Deterministic Problems in the Natural Sciences*, MacMillan (New York).

Lindzen, R. S. 1990: *Dynamics in Atmospheric Physics*, Cambridge University Press (Cambridge, UK), Chapter 2.

Livio, M., 2002: *The Golden Ratio: The story of Phi, the World's Most Astonishing Number*, Broadway Books (New York).

Lorenz, E. N., 1963: Deterministic nonperiodic flow, *J. Atmos. Sci.*, **20**, 130–141.

Lorenz, E. N., 1993: *The Essence of Chaos*, University of Washington Press (Seattle).

Manabe, S., and R. T. Wetherald, 1967: Thermal equilibrium of the atmosphere with a given distribution of relative humidity, *J. Atmos. Sci.*, **24**, 241–259.

Markowsky, G., 1992: Misconceptions about the Golden Ratio, *Coll. Math. J.*, **23**, 2–19.

Marsili, M., and Y.-C. Zhang, 1998: Interacting individuals leading to Zipf's law, *Phys. Rev. Lett.*, **80**, 2741–2744.

May, R., 1974: Biological populations with non-overlapping generations: stable points, stable cycles, and chaos, *Science*, **186**, 645–647.

McKenna, P. J., 1999: Large torsional oscillations in suspension bridges revisited: Fixing an old approximation, *Am. Math. Monthly*, **106**, 1–18.

Murray, J. D., 2002: *Mathematical Biology, I: An Introduction*, Springer-Verlag (New York).

Myers, R. A., N. J. Barrowman, J. A. Hutchings, and A. A. Rosenberg, 1995: Population dynamics of exploited fish stock at low population levels, *Science*, **269**, 1106–1108.

Newman, M. E. J., 2003: The structure and function of complex networks, *SIAM Rev.*, **45**, 167–256.

North, G. R., 1975: Analytical solution to a simple climate model with diffusive heat transport, *J. Atmos. Sci.*, **32**, 1301–1307.

Pearl, R., and L. J. Reed, 1920: On the rate of growth of the population of the United States since 1790 and its mathematical representation, *Proc. Natl. Acad. Sci. U.S.A.*, **6**, 275–288.

Penland, C. L., and P. D. Sardeshmukh, 1995: The optimal growth of tropical sea surface temperatures, *J. Climate*, **8**, 1999–2024.

Perelson, A. S., A. U. Neumann, M. Markowitz, J. M. Leonard, and D. D. Ho, 1996: HIV-1 dynamics in vivo: Virion clearance rate, infected cell lifespan, and viral generation time, *Science*, **271**, 1582–1586.

Perelson, A. S., and P. W. Nelson, 1999: Mathematical analysis of HIV-I: Dynamics in vivo, *SIAM Rev.*, **41**, 3–44.

Pierrehumbert, R. T., 2005: Climate dynamics of a hard snowball Earth, *J. Geophys. Res.*, **110**, DO1111, doi: 10.1029/2004JDO05162.

Price, D. de S., 1965: Networks of scientific papers, *Science*, **149**, 510–515.

Price, D. de S., 1976: A general theory of bibliometric and other cumulative advantage processes, *J. Am. Soc. Inf. Sci.*, **27**, 292–306.

Rayleigh, Lord (R. J. Strutt), 1916: On convective currents in a horizontal layer of fluid when the higher temperature is on the underside, *Philos. Mag.*, **32**, 529–546.

Saltzman, B., 1962: Finite amplitude free convection as an initial value problem-I, *J. Atmos. Sci.*, **19**, 329–341.

Schmidt-Nielsen, K., 1984: *Scaling: Why Is Animal Size so Important?* Cambridge University Press (Cambridge, UK).

Sellers, W. D., 1969: A climate model based on the energy balance of the earth–atmosphere system, *J. Appl. Meteorol.*, **8**, 392–400.

Shipman, P. D., and A. C. Newell, 2004: Phyllotactic patterns and plants, *Phys. Rev. Lett.*, **92**, 168102-1–168102-4.

Strogatz, S., 1994: *Nonlinear Dynamics and Chaos*, Westview Press (Boulder, CO).

Strogatz, S. H., 1988: Love affairs and differential equations, *Math. Mag.*, **61**, 35.

Strutt, R. J., 1906: On the distribution of radium in the Earth's crust, and the Earth's internal heat, *Proc. R. Soc. London* A, **77**, 472–485.

Taylor, J., 1859: *The Great Pyramid: Why Was It Built and Who Built It?* Longman, Green, Longman, and Roberts (London).

Thompson, C., and D. Battisti, 2000, 2001: A linear stochastic dynamical model of ENSO, Part I: Development. *J. Climate* **13**, 2818–2883. Part II: Analysis, *J. Climate*, **14**, 445–466.

Thompson, D. W., 1917: *On Growth and Form*, republished in 1992 by Dover Publications (New York).

Thomson, W. (Lord Kelvin), 1864: On the secular cooling of the Earth, *Trans. R. Soc., Edinburg*, **23**, 167–169.

Thomson, W. (Lord Kelvin), 1866: The "Doctrine of Uniformity" in geology briefly reputed, *Proc. R. Soc. Edinburgh*, **5**, 512–513.

Vallis, G. K., 1986: El Niño: A chaotic dynamical system? *Science*, **232**, 243–245.

Vallis, G. K., 1988: Conceptual model of El Niño and the Southern Oscillation, *J. Geophys. Res.*, **93**, 13979–13991.

Walker, G., 2003: *Snowball Earth: The Story of a Maverick Scientist and His Theory of the Great Global Catastrophe that Spawned Life as We Know It*, Crown Publishers (New York).

Wan, F. Y. M., 1989: *Mathematical Models and Their Analysis*, Harper & Row (New York).

West, J. B., J. H. Brown, and B. J. Enquist, 1997: A general model for the origin of allometric scaling laws in biology, *Science*, **276**, 122–126.

West, J. B., J. H. Brown, and B. J. Enquist, 1999: The fourth dimension of life: Fractal geometry and allometric scaling of organisms, *Science*, **284**, 1677–1679.

West, J. B., J. H. Brown, and B. J. Enquist, 2001: A general model for ontogenetic growth, *Nature*, **413**, 628–631.

Wroe, S., 1999, *News in Science*, April 6.

Yorke, J., and T.-Y. Li, 1975: Period three implies chaos, *Am. Math. Monthly*, **82**, 985–992.

Zebiak, S. E., 1984: *Tropical Atmosphere–Ocean Interaction and the El Niño/Southern Oscillation Phenomenon*, Ph.D. Thesis, MIT (Cambridge, MA).

Zebiak, S. E., and M. A. Cane, 1987: A model El Niño-Southern Oscillation, *Monthly Weather Rev.*, **115**, 2267–2278.

Index